UNDERSTANDING PRESSES AND PRESS OPERATIONS

Donald F. Wilhelm
Editor
Published by:
Society of Manufacturing Engineers
Marketing Services Department
One SME Drive
P.O. Box 930
Dearborn, Michigan 48128

UNDERSTANDING
PRESSES
AND PRESS OPERATIONS

Library of Congress Catalog Card Number: 81-51805

International Standard Book Number: 0-87263-069-2

Manufactured in the United States of America

SME wishes to express its acknowledgement and appreciation to the following publications for supplying the various articles reprinted within the contents of this book.

Machine and Tool Blue Book
Hitchcock Publishing Company
Hitchcock Building
Wheaton, IL 60187

Metal Stamping
American Metal Stamping Association
27027 Chardon Road
Richmond Heights, OH 44143

Modern Machine Shop
600 Main Street
Cincinnati, OH 45202

Precision Metal
A Penton/IPC Publication
614 Superior Avenue West
Cleveland, OH 44113

Production Engineering
A Penton/IPC Publication
Penton Plaza
Cleveland, OH 44114

Grateful acknowledgement is also expressed to:

Mr. F.E. Heiberger
Danly Machine Corporation
2100 South Laramie Avenue
Chicago, IL 60650

Mr. Robert Lown
Greenerd Press and Machine Company Inc.
41 Crown Street
Nashua, NH 03061

Mr. Thomas Marquardt
Komatsu America
35522 Industrial Road
Livonia, MI 48150

Mr. William B. Ardis
The Minster Machine Company
240 West Fifth Street
Minster, OH 45865

Mr. George Petruck
Power Gems, Inc.
2425 Devon Avenue
Elk Grove Village, IL 60007

Mr. T.M. Motta
Mr. Paul Kjelstrom
The Verson Allsteel Press Company
1355 East Ninety-third Street
Chicago, IL 60619

Mr. Sheldon E. Young
Vibro Dynamics Corporation
500 East Plainfield Road
P.O. Box 188
Countryside, IL 60525

PREFACE

Today's mechanical press permits the economical manufacture of mass-produced parts and is a primary contributor to our high standard of living. Every segment of today's modern life uses parts produced on dies utilizing presses. And evidence shows increasing popularity of the press in production applications. Mechanical power presses are one of industry's most important tools—over 300,000 are in use in the United States today. As a result, current information on presses is vital to everyone involved in manufacturing planning or operations.

Understanding Presses And Press Operations explores many of the facets of the modern-day press. It covers the information you should know about presses from the principles to selection, replacement or rebuilding, costs, noise, repair and the type of press that will be in factories tomorrow. The book will help you with the most efficient control and use of plant equipment.

The history of the modern press, which began in the 1850's, is expertly explained by Tom Marquardt in the first paper reprinted in this book. This paper is one of two readings in Chapter One, Principles. In the second paper, Nick Fisher presents a discussion intended to acquaint the engineer and general company management with the principles of mechanical power presses.

Many factors should be considered in the planned purchase of a metal forming press. Chapter Two, "Selection", contains the presentation "Considerations For Selecting a Press", which is followed by information on selecting a press for the transfer die and a paper discussing total system, heavy-tonnage, high-production transfer presses.

Chapter Three, "Installation, Maintenance and Repair", presents information on preventative maintenance and the proper mounting and isolation of presses. If a plant has many power presses, the chances are good that there is a noise problem. This volume contains two articles which offer suggestions in noise reduction.

Chapter Four, "Stamping Process", discusses hydraulic overloads, measuring loads in the metalworking process and snap-through arrestors for high-speed blanking. Chapter Five, "New Strides In Productivity", provides a look ahead. Some of the press advances that are today spurring productivity are explored. The chapter details the characteristics of hydraulic presses and looks at the future of these machines.

This book's Appendix, "Know Your Presses", is meant to supplement the material in this volume by providing other examples of presses and equipment. Class or clinic instructors will also find this section useful in providing illustrations.

I wish to thank the authors of each of these articles for their contributions. (Titles are those that they held when they wrote the journal article or paper.) I also wish to express my gratitude to the publications who supplied some of the material in this volume. They include: **Machine and Tool Blue Book, Metal Stamping, Modern Machine Shop, Precision Metal** and **Production Engineering.** In addition, special thanks to some of the press companies for supplying material for this book. They include: E.W. Bliss, the Danley Machine Corporation, Greenerd Press and Machine Company Inc., Komatsu America, the Minster Machine Company, Power Gems Inc., the Verson Allsteel Press Company and Vibro Dynamics Corporation. In addition, my thanks to Bob King and Judy Stranahan of the SME Marketing Services Department for their help in assembling this volume.

Donald F. Wilhelm
President
Helm Instrument Company Inc.
Editor

SME

The informative volumes of the Manufacturing Update Series are part of the Society of Manufacturing Engineers' effort to keep its Members better informed on the latest trends and developments in engineering.

With 60,000 members, SME provides a common ground for engineers and managers to share ideas, information and accomplishments.

An overwhelming mass of available information requires engineers to be concerned about keeping up-to-date, in other words, continuing education. SME Members can take advantage of numerous opportunities, in addition to the books of the Manufacturing Update Series, to fulfill their continuing educational goals. These opportunities include:

- Chapter programs through the over 200 chapters which provide SME Members with a foundation for involvement and participation.

- Educational programs including seminars, clinics, programmed learning courses and videotapes.

- Conferences and expositions which enable engineers to see, compare, and consider the newest manufacturing equipment and technology.

- Publications including Manufacturing Engineering, the SME Newsletter, Technical Digest and a wide variety of books including the Tool and Manufacturing Engineers Handbook.

- SME's Manufacturing Engineering Certification Institute formally recognizes manufacturing engineers and technologists for their technical expertise and knowledge acquired through years of experience.

In addition, the Society works continuously with the American National Standards Institute, the International Standards Organization and other organizations to establish the highest possible standards in the field.

SME Members have discovered that their membership broadens their knowledge throughout their career.

In a very real sense, it makes SME the leader in disseminating and publishing technical information for the manufacturing engineer.

TABLE OF CONTENTS

CHAPTERS

TABLE OF CONTENTS (Cont.)

4 CONTROL OF THE STAMPING PROCESS

5 NEW STRIDES IN PRODUCTIVITY

APPENDIX

INDEX

CHAPTER 1

PRINCIPLES

Prepared for *Understanding Presses And Press Operations*

Advances In Press Technology

By Thomas Marquardt
General Manager
Industrial Machinery Division
Komatsu America Corp.

Illustrations courtesy of E.W. Bliss Division, Gulf & Western Manufacturing Company

The beginning of metal forming evolved from the blacksmith era of manual hammer forming of metal. These meager beginnings (Figure 1 and 2) have grown to the current position of over 5,000 companies involved primarily in the production of stamped metal parts.

Figure 1

Figure 2

The Inca Indians, coining metal on a wooden screw press, were among the first innovators of the screw press. From these early beginnings, the style of machine shown in Figure 3 emerged in the 1850's.

Figure 3

The machine (Figure 3) utilized the driven power of the line shaft to the flywheel of the press. The first offering contained no clutch and ran continuously. Trying to time the work entry into a continuously running press was very difficult, which led to the development of a mechanical clutch to permit intermittent tripping of the press. This permitted engaging and disengaging of the clutch tripping, (Figure 4 and 5) but not single tripping as we know it today. Late in the 1880's, a new mechanical friction clutch was introduced. This clutch used a series of wooden blocks for a

Figure 4

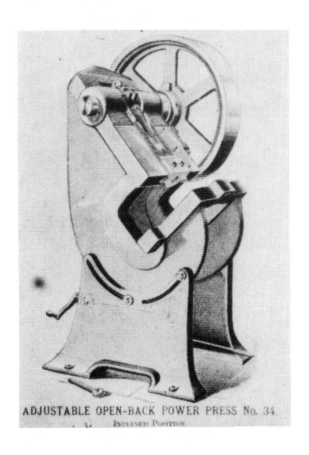

ADJUSTABLE OPEN-BACK POWER PRESS No. 34.
INCLINED POSITION.

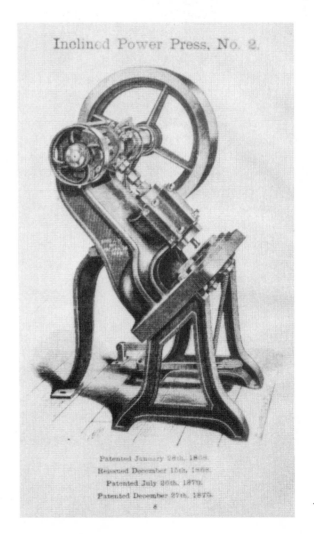

Inclined Power Press, No. 2.

Patented January 26th, 1864.
Reissued December 15th, 1868.
Patented July 26th, 1870.
Patented December 27th, 1870.
6

friction lining and was the forerunner of the air friction
clutches used today.

Figure 5

Figure 6

As metal part shapes became more sophisticated, different
type machines to produce these shapes became necessary.
Figure 6 illustrates a double action press with blankholder
and plunger for large and small draw requirements.
As the metal frame sizes and bodies for automobiles grew,
the first large press requirements emerged. Next to be de-

veloped was early automation attachments to improve the standard of work measurement.

A series of twin end drive toggle presses (Figure 7) were introduced in April of 1910 to meet the continuing needs of industry. This machine was used by Chalmers Motor Company to manufacture drawn fenders. As automotive production grew and production rates increased to supply the assembly lines, so did the need for duplicate parts. The demand for material handling and automatic feeds also increased the demand for duplicate parts.

Figure 7

The history of the press reflects the need for new machine tools which have the ability to furnish intricate machined parts in order to supply other machine tools. As a result of this need, high-speed, single-point, gap frame presses and two-point, high-speed, straight side presses were developed.

The introduction of new items has helped industry meet the challenge of increased legislation and rising costs. The use of sliding bolsters to carry a part into the tool area, prevents the operator from entering the hazard area - a result of government safety regulations. Other examples include:

- New types of guarding (barrier, light curtains).
- Hydraulic overload mechanisms.
- Strain sensing equipment to protect tooling.
- New clutches that permit higher single trip rates and faster stopping times.
- Retrofits of older equipment with new air friction clutches. (Figure 8).

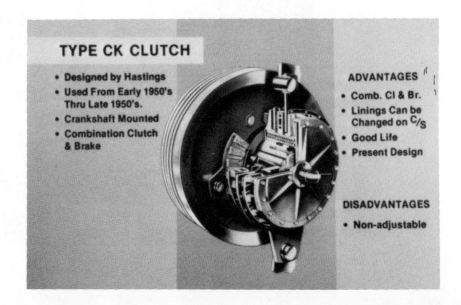

Figure 8

However, metal stamping continues to be one of the most economical methods of producing a part. This fact is demonstrated by rotary redraw presses that manufacture brass tubes at production rates of 1,200 parts per minute. The hot forging machine (Figure 9) replaces four machines, thus conserving floor space and increasing productivity. This is important considering that the stroke of the press is 48 inches and a hydraulic extender unit on the last station of the tooling, increases the effective stroke to 72 inches.

Figure 9

The future challenges to the manufacturing engineer include:

. Design tools to guarantee the safety of the operator.

. Efficient control of the use of plant equipment.

. Design for efficient means of manufacturing.

Additional factors facing the manufacturing engineer includes some of the following points.

. The U.S. needs over 4,000 experienced tool and die makers per year to meet the retirement rate. In 1980, only 1,300 tool and die makers met that challenge.

. Experienced machinists must be trained at a rate of 7,500 per year to meet the retirement rate. In 1980, only 2,800 people were trained to meet that need.

. With the average age of the skilled labor force being 57-years-old, (60 percent of the total skilled labor force will retire in six to eight years) the training requirement for skilled personnel must be recognized.

Presented at the 1976 International Tool and Manufacturing Engineering Conference, April, 1976

Principles Of Mechanical Power Presses

By Nicholas Fisher
Vice President—Sales
L & J Press Corporation

The mechanical power press permits the economic manufacture of mass produced parts which is a primary contributor to our high standard of living. Every segment of life uses parts produced on dies utilizing presses such as automobiles, planes, appliances, hardware, telephones, beverage containers, computers, jewelry and many many more. This discussion is intended to acquaint the engineer and general company management with the principles of mechanical power presses.

What is a mechanical power press?

A mechanical power press is a machine used to supply power to a die that is used to blank, form, or shape metal or non-metallic material. Thus, a press is a component of a manufacturing system that combines the press, a die, material and feeding method to produce a part. The designer of the manufacturing system which incorporates a press, die, feeding method, and material must also provide proper point of operation guards. While each of the components of this manufacturing system is important, this discussion will highlight the mechanical power press.

Types of presses:

There are over 300,000 presses in use in the United States; such as Double Action, Triple Action, Knuckle Joint, Horizontal, Adjustable Bed, Horn, Open Back Inclinable, Gap, Double Crank, Single Crank, Four Point, Extrusion, High Speed, Slow Speed, and many other specialized type presses that need not be mentioned in an introductory discussion. Because we are considering basics, we will limit our discussion to Single Action presses, those that have one slide and one actuating motion. In this group there are two predominant press types; Open Back Inclinable and Straight Sides which we will consider in this discussion.

Press drive types:

A. Non-Geared (flywheel type)

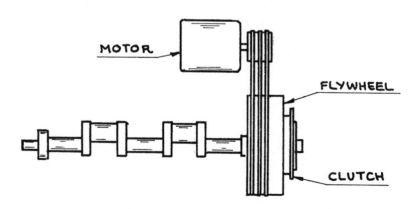

FIGURE 1

B. Single Geared single end drive

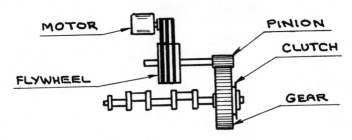

FIGURE 2

C. Single Geared twin end drive

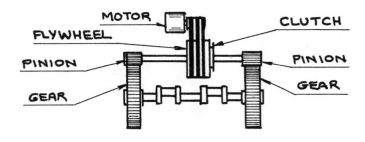

FIGURE 3

D. Double Geared twin end drive

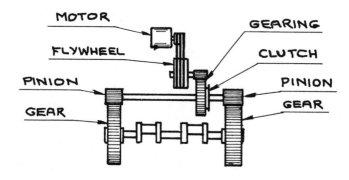

FIGURE 4

In non-geared presses the flywheel is located on the crankshaft and driven directly from a motor through a belt drive.
Non-geared presses are used primarily for blanking and some
shallow forming operations. Because they do not have a gear
reduction, they are used for high speed stamping operations.
Special flywheel presses, depending upon size and features
included, can operate in ranges from 90 and in some cases
over a 1,000 strokes per minute. Also, because there are no
gears and the flywheel is located on the crankshaft, the
tonnage is rated at close to the bottom of the press stroke.

In single geared presses the flywheel is mounted on the backshaft and the power is then transmitted through a pinion to
a main gear mounted on the crankshaft. Single geared presses
operate in the speed range of 35 to 100 strokes per minute
and in some cases with special gearing arranged to operate up
to 120 or 150 strokes per minute. Because these presses utilize a gear reduction, with the flywheel on the high speed
backshaft, more flywheel energy can be provided than in a
non-geared press; thus making the single geared press better

suited for drawing and heavy forming operations.

Double geared presses use two gear reductions from the flywheel to the crankshaft achieving a speed range normally from 8 to 20 strokes per minute. These presses are used for the more difficult heavy deep drawing operations.

Press component identification:
The following drawings identify by proper names the principle press components for O.B.I. and Straight Side Presses.

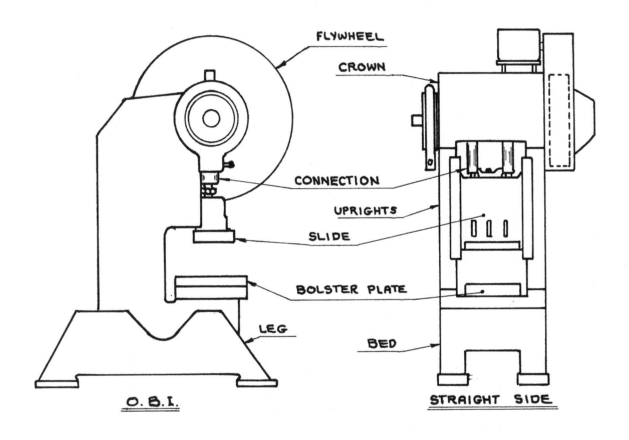

FIGURE 5

Shutheight:
The space available between the press bolster and the slide is the work area, or die space of the press. This distance is called the shutheight and is identified two ways.

  A. Shutheight from top of bed to face of slide, stroke down adjustment up.
  B. Shutheight from top of bolster to face of slide, stroke down adjustment up.

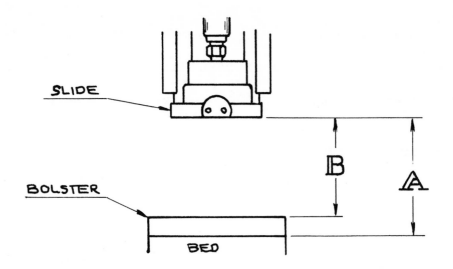

SLIDE

BOLSTER

BED

FIGURE 6

Sample tonnage calculation:
A mechanical press is one component in a manufacturing system and is useless without a die and material to put into it. Also, the size of the part which is related to how much tonnage or work will be done in the press, determines the size of the press to be used on a particular application. The determination of a calculated tonnage to produce a part is a subject in itself and we will not cover it in this discussion; but, as an elementary example let us look at the basic blanking formula:

Pressure in tons = length of cut X material thickness
X $\frac{\text{shear strength}}{2,000 \text{ lbs.}}$

Note:　A.　1.　No shear on die ------Add 25% to tonnage.
　　　　　　2.　Partial shear on die--Full tonnage.
　　　　　　3.　Full shear on die-----Reduce tonnage by 25%.

Note:　B.　Shear on die: Construction of a die (usual blanking) in such a way all the work is not done at once. The die is designed so that the cut edge will vary around the blank (similar to the action of scissors cutting paper).

　　　　　　For illustration, consider a 6" diameter piece of steel to be blanked from mild steel with a 50,000 PSI shear strength, .065" thick with no shear on the die.

　　　　　　(6 X 3.1416) X .065 X $\frac{50,000}{2,000}$ = 30.6 tons

　　　　　　Add 25% for No Shear----------　7.6 tons
　　　　　　　　　　　　　TOTAL　　38.2 Tons

Thus, this job could be considered theoretically for a 40 ton non-geared press; but, since most press builders do not make a 40 ton press a 45 ton model would most likely be considered.

Let us consider three blanking operations of different thickness and diameter.

1. 25" Diameter Blank
    .025" Thick
    Mild Steel, 50,000 PSI Shear Strength
    Partial shear on die

    $$(25 \times 3.1416) \times .025 \times \frac{50,000}{2,000} = 50 \text{ Tons}$$

2. 10" Diameter Blank
    .062" Thick
    Mild Steel, 50,000 PSI Shear Strength
    Partial shear on die

    $$(10 \times 3.1416) \times .062 \times \frac{50,000}{2,000} = 50 \text{ Tons}$$

3. 5" Diameter Blank
    .125" Thick
    Mild Steel, 50,000 PSI Shear Strength
    Partial shear on die

    $$(5 \times 3.1416) \times .125 \times \frac{50,000}{2,000} = 50 \text{ Tons}$$

We can see each of the above requires 50 tons but because of the varying thickness between the different blanks the energy required will vary by the distance through which the work is performed. Although in actual practice the distance through which work is performed would probably be one half of metal thickness due to the blank breaking through after the punch has penetrated half the thickness. But, we will use full thickness for our example energy calculation.

1. 50 tons X .025" = 1.25 inch tons.

2. 50 tons X .062" = 3.10 inch tons.

3. 50 tons X .125" = 6.25 inch tons.

As can be seen, even though all three jobs require 50 tons, the energy required increases due to the greater distance through which the work must be done.

When considering a press to perform the above, we would look at a 60 ton press as most press builders do not manufacture a

50 ton press. We also must consider at what distance above bottom the press should be rated to do 60 tons of work because of the various energy requirements of the jobs. We know most builders rate 60 ton presses as follows:

> 60 ton non-geared press rated .0625" distance of above bottom.

> 60 ton geared press rated .250" distance of above bottom.

Thus, in our example the 25" diameter blank and the 10" diameter blank could be assigned to a 60 ton non-geared press while the 5" diameter blank should be considered for a 60 ton geared type press.

Another factor to consider would be the bed and slide area. Most 60 ton O.B.I. presses have a bed and slide area of 32" right to left X 18" front to back, which would be large enough for the 10" and 5" diameter blanks; but, too small for the 25" diameter blank. Thus, a 60 ton press with a wide bed should be considered, and if this is not available possibly a larger tonnage press with sufficient bed and slide area to handle the larger die.

We have seen an example of determining job tonnage, let us now look farther in to the considerations a press builder looks at in determining press capacity.

Determination of press capacity:
After an estimate is made to what tonnage is necessary, a press has to be selected with proper specifications to match the tonnage required of the die. The press builder normally designs his press considering the following main criteria:

> A. Component Strength

> B. Flywheel Energy

> C. Torque Capacity

(A) Component strength is the physical strength of the various parts of the press necessary to withstand the load and resist deflection within designed tolerances.

(B) The motor furnishes the energy to the flywheel and the flywheel stores the energy until a certain amount of energy is removed from the flywheel by each working press stroke necessary to perform work.

(C) The torque capacity is the ability to take the energy of the flywheel and transmit it through the gears (if a geared press), clutch, crankshaft, connection and slide into the die and into the work.

Motor drives:

While there are many motor drives available, such as DC, two speed, gear head, constant energy, etc.; the three basic drives used on presses are:

    A. Single Speed

    B. Mechanical Variable Speed

    C. Eddy Current

(A) The single speed motor powers the press at a constant fixed speed and for all practical purposes never changes output speed.

(B) The mechanical variable speed motor drive is a package unit containing a single speed motor with a belt and movable sheave arrangement that changes the speed of the drives output shaft. These drives are normally supplied in speed ranges of 2:1, 3:1 and in some cases 4:1.

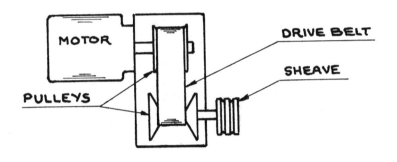

FIGURE 7

(C) The eddy current motor drive is a package unit containing a single speed motor and an eddy current coupling that changes the drive's output shaft speed. While the eddy current drive speed ratio is theoretically infinite their effective range is normally in the 3:1, 4:1, or 5:1 speed range.

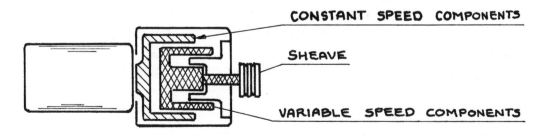

FIGURE 8

Why use a variable speed drive?
A variable speed drive permits the stamper to adjust the
press speed for optimum performance taking into consideration
the die, feed, part ejection and production requirements.

Flywheel energy versus press speed:
The press user must work close with the press builder in
determining the rating of flywheel energy for the variable
speed drive. One must keep in mind that flywheel energy is
reduced (not press tonnage capacity) when the speed is reduced below its standard rating. The flywheel energy is reduced by the square of the speed reduction. Conversely if
the speed is increased the flywheel energy is increased by
the square of the speed increase.

As an example, lets consider a theoretical 60 ton non-geared
O.B.I. press with a variable speed drive arranged for 50 to
200 strokes per minute, providing 60 tons through 1/16" distance above bottom at a rated speed of 100 strokes per minute.
If the speed is doubled to 200 strokes per minute, you will
theoretically have four times the energy; but, because this
press is designed with component strength and torque capacity
to 60 tons, the press is still rated at 60 tons at 1/16"
distance above bottom for any speed above the standard rating.
If the speed is reduced, by 1/2 to 50 strokes per minute, the
energy is actually reduced by 3/4 because at 1/2 speed there
is only 1/4 the energy left in the flywheel. The 60 ton
press thus would supply 15 tons through 1/16" or 60 tons
through 1/64". At 50 strokes per minute the press is still a
60 ton press from a component strength and torque capacity
standpoint, but the flywheel energy is reduced to 1/4 the
standard rating; thus its ability to do work at the 50 stroke
per minute speed is greatly reduced.

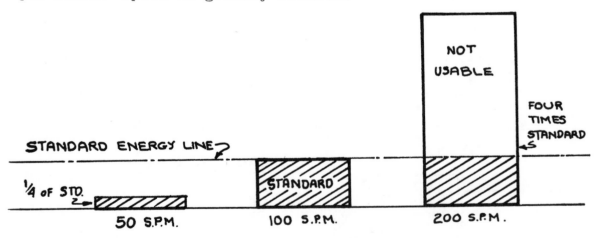

FIGURE 9
If the speed range is prudently considered in selecting a
press and the press applied properly, this loss of energy at
slower speeds should not present a problem to the stamper.

Lubrication:
Like any piece of machinery proper lubrication is essential
to overall performance and long life. There are three basic
types of lubrication systems normally offered on mechanical
power presses.

    1. Manual Hand Pump Or One Shot System

    2. Automatic Or Centralized (lost oil)

    3. Recirculating

The manual hand pump systems are usually used on smaller
presses and those running in single stroke applications.

Automatic systems usually utilize a reservoir and a means of
automatically pumping oil through the lubrication system such
as air pressure. The amount of oil flow can be controlled by
meters, valves, timers or the lubrication fitting itself.
These types of systems while automatically sending oil
through the system are of a lost oil type and thus require
the oil reservoir to be manually filled as oil is used.

The recirculating oil systems are the most sophisticated in
that they continuously pump oil from a reservoir through the
system collecting the oil in catchers and drip pans and
direct the oil back into the oil reservoir, where it can be
filtered and again pumped through the lubrication system.
Recirculating oil is usually used on high speed press applications and automatic straight side presses.

Clutch:
The clutch is a mechanism used to control the coupling of the
flywheel (or gear on a greared press) with the press crankshaft.

While many older presses utilize a mechanical clutch, which is
a full revolution type; meaning when the clutch is activated
it cannot be disengaged until the crankshaft has made one complete revolution. Most modern presses are equipped with an
air friction clutch and brake arrangement that is commonly
called a part revolution clutch. The part revolution clutch
can be disengaged at any point in the stroke before the crankshaft has completed a full revolution. The air friction
clutch permits versitility of press use such as; increased
press speeds, controls can be added to protect valuable dies
because of the part revolution stopping characteristic, and
less down time as air clutch presses tend to provide longer
maintenance free operation.

Air counterbalance:
Air counterbalances are air cylinders mounted in the press
frames and connected by the cylinder rod or rod brackets to

the press slide. Air pressure is placed in the counterbalance cylinder that tends to lift the slide to a neutral
weight condition. As die weight is added to the slide,
additional air pressure is added to the air counterbalance
to compensate for this added weight. Air counterbalancing
aids in the overall performance of a press as follows:

    A.   To counterbalance the weight in the slide and
        the die components that are attached to the slide,
        assisting in providing a smooth press stroke.

    B.   Proper air counterbalance is used to take up bearing
        clearances before the die closes and work is per-
        formed.

    C.   Proper air counterbalancing tends to assist in
        stopping the press because less load is applied
        to the brake.

    D.   Proper air counterbalancing helps remove backlash
        caused by tooth clearance in press gearing.

Automatic and high speed presses:
High speed is a very relative term, to a user of a 1,000 ton
press who normally runs his press at 10 to 12 strokes per
minute a speed of 20 strokes per minute would be high speed;
at the other end of the spectrum there are some applications
that a 30 to 60 ton press can run 1,500 to 1,800 strokes per
minute, but, these are special cases and only a very few parts
fall into a category that would require a speed in the range
of the above two examples. For our discussion utilizing
presses 200 tons and under we see high speed for geared type
presses use usually somewhere between 60 and 120 strokes per
minute and in non-geared presses anywhere between 300 and 800
strokes per minute. Most parts, because of their configuration, and many dies because of their complexity and cost
require press speeds in a more conservative range permitting
long trouble free runs with very little down time.

Conventional O.B.I. and Gap presses can be used in some instances to run progressive or multiple station dies if the
load in the die is balanced. But, if at all possible, to
achieve longer die life, better part quality, and improve
press performance a double crank straight side press is recommended for progressive die operations. Automatic production is achieved by feeding the material into the die by a
feed, either independently driven such as an air feed or
press driven such as a roll feed. Automatic production using
progressive dies permits operations that were previously single stroked in separate presses to be combined in one die and
one press and usually performed at a higher rate of speed.
Thus, the cost per piece is usually reduced because less labor
per part is required and stock saving is achieved because coil

stock is used instead of strip or precut blanks.

Regardless if the press is a gap type having a fixed base for
rigidity or a double crank straight side press, the following
features are necessary for the press to best run high speed
automatic production.

   1. Frame designed to reduce deflection and absorb shock
       of stamping load.

   2. Crankshaft counterweighted for smooth high speed
       operation.

   3. Air counterbalance.

   4. Gibs utilizing bronze, non-metallic liners, or the
       new Teflon bronze liners enabling high velocity
       slide travel and close gib setting to reduce toler-
       ances.

   5. Recirculating oil lubrication.

   6. Variable speed motor drive.

   7. Slide and bed of adequate size to contain die.

   8. Adequate feed system.

Examples of an automatic straight side and an automatic gap
frame press are shown in photograph "A" and "B" at the back
of this paper.

Press feeds:
While this discussion is centered along mechanical power
presses, the brief mention of press feeding as feeds are of
a necessity when considering an automatic or high speed press
application. There are many types of feeds that can be used
in combination with a press, the most common are: air feeds,
slide feeds, roll feeds, and cam feeds. All but air feeds
are usually mounted on the press and driven from extension on
the press crankshaft. Care must be used when considering a
feed mounted on the press to make sure the feed will handle
the material to be used and will feed in the proper direction.
First, the limit of material capacity is usually stated as
follows:

      Material type:
      Maximum Width ____ at ____ inches thickness
      Maximum Thickness ____ at ____ inches width
      Maximum feed length _____ inches
      Feed line height over bolster _____ inches
      Feed line height adjustment plus or minus ____ inches
      Direction of stock flow: Right to Left or Left to
        Right

Summary:
A press can be a very simple machine or a very complex piece
of production equipment depending upon how sophisticated the
manufacturing system the press is used in has been designed.
By starting from a basic understanding of the principles of
mechanical presses an engineer can open up new uses for this
equipment that is so vital to todays high production needs.

PHOTOGRAPH "A"
Automatic Straight Side Press

Photo used for illustration only.
Does not include all guarding as
required by OSHA.

23

PHOTOGRAPH "B"
Automatic High Speed Gap Frame Press

Photo used for illustration only.
Does not include all guarding as
required by OSHA.

CHAPTER 2

SELECTION

Considerations For Selecting A Press

Many factors should be given serious consideration in the planned purchase
of a mechanical metal forming press. Making the proper decisions in these
areas can improve part quality, extend die life, increase production speeds
and reduce press downtime. In many cases the "proper decision" simply means
selecting the press that most closely fits the work you intend for it to
do ... matching the press to the job.

This paper is designed to point out many of the major elements involved in
selecting the correct press for your needs ... and why these decisions are
important to a successful operation. There are, of course, many additional
important factors to consider in selecting a press to meet your needs. Working closely with press suppliers and letting them know your needs and application plans for the equipment is your best assurance of success.

The basic construction of the straight side press consists of four main
parts: a bed [#1], two uprights [#2 and #3], and the crown [#4]. (Figure 1).
These frame components are secured in a loaded position by means of four tie
rods which are the vertical members. Upon assembly of these components the
tie rods are heated so that, when cool, the contracted force of the rods pre-
loads the press to approximately a load twice the rated capacity of the machine. The press, then, is held together by tension due to the elongation
of the tie rods.

Figure 1

The left illustration in Figure 2 shows a heated rod which elongates. At this point the nuts on each end of the rod are tightened and when the rod is allowed to cool, as shown by the illustration on the right, the forces in the rod produce tension so that the machine components are held together in compression.

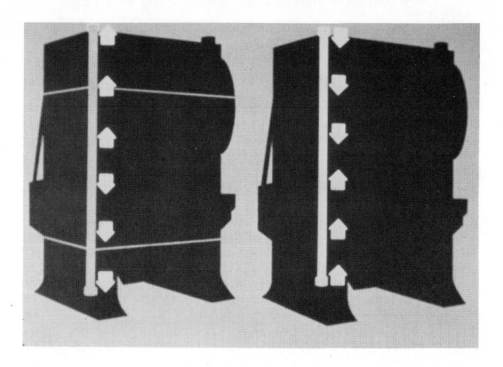

Figure 2

The illustration shown on the left in Figure 3 displays what happens when a machine is improperly designed to hold the force exerted by the tie rods. The illustration on the right shows the principle of good design. The tie rod does not excessively compress the frame members, and as a result, the machine is extremely rigid because press components do not elongate when a load is applied.

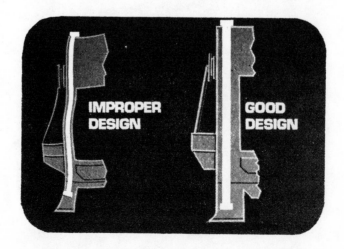

Figure 3

Figure 4 shows two machines, both rated at 100 tons capacity. The machine at the left is improperly designed. The illustration at the right shows the mass that is required in proper design to achieve good die life. This principle must be considered as fundamental for the proper application of presses used with progressive dies where die life becomes extremely important.

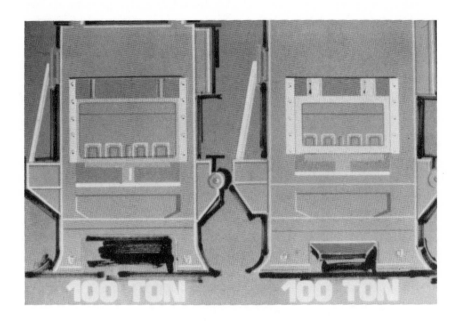

Figure 4

When a machine is improperly designed and cannot contain the force exerted by the tie rods, the punch has to enter the die excessively as shown by the illustration on the left in Figure 5. The result is excessive die wear. On the right, notice that the punch enters the die only far enough to produce the proper piercing depth and drive the slug into the die. The result is much less side wear on the punch.

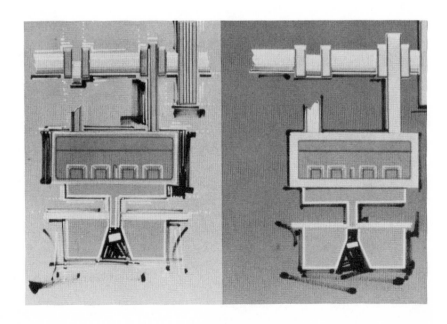

Figure 5

Rigidity in the machine is not enough. Look at this die arrangement in
Figure 6. The die must be properly applied to the correct bed opening in
the machine. All too frequently we find the bed opening of the machine is
far in excess of what it should be for proper die support. This illustration shows that only the bolster plate is supporting the die. Look at what
happens because of the bed opening under the bolster. The result of such
an application is that die life is cut to one-fourth or less of what its
normal wear should be.

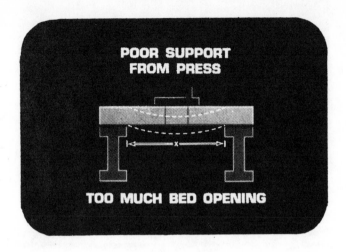

Figure 6

All too frequently, a die is not large enough to fit in the press. Figure 7
simplifies the arrangement, but the purpose here is to show what is hidden
beneath a bolster plate which covers up these misapplications. Notice that
the die is not only receiving extremely poor support from the press bed but
it is also improperly located.

Figure 7

Figure 8 illustrates a properly loaded arrangement. The press is built
with the proper bed opening for the die. In this way, the die receives
support entirely around the die shoe. The opening is only of sufficient
width and length to take care of part removal, and die shoe deflection is
minimal.

Figure 8

Off-center loading is sometimes impossible to avoid. Frequently modern die
designs require off-center load characteristics in order to incorporate
many of the additional operations demanded by present progressive dies. It
is important to consider proper press design when off-center loading is a
necessity. The press selected should be one which is designed to handle
off-center loads. A properly designed progressive die press is capable of
handling off-center loads equal to two-thirds the rated capacity of the
machine without greater deflection than if the full capacity of the machine
were placed across the center area of the ram. Figure 9 exaggerates how an
off-center load may force the press slide to rotate out of parallel.

Figure 9

Figure 10 illustrates what happens when the press is not properly designed for the use of tie rod construction. Spongey construction such as this would cause the machine to breathe excessively.

Figure 10

The machine having two connections and double throw crank deflects much less under a load that is off the center of the press, as shown in Figure 11.

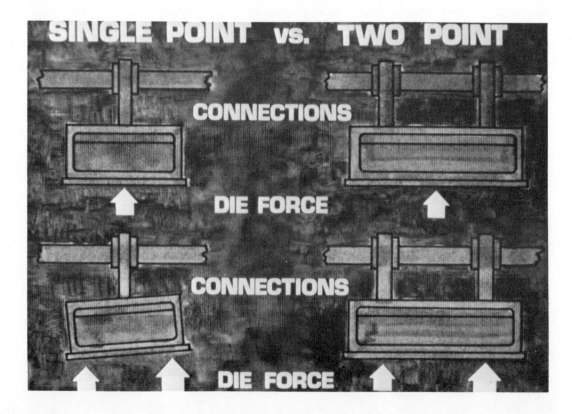

Figure 11

When a die is positioned too far forward as shown in Figure 12, the net result is that the gibs of the press must carry the off-center load. This is a poor situation, because on most progressive die presses the gibs are not designed for extreme off-center loads in a front to back direction. Note, that off-center loading in a right to left direction and with a two point design machine, is not nearly as disastrous as front to back off-center loading, because the crank can help restrain the off-center load.

Figure 12

In Figure 13, the die is simply positioned too far to the rear. While these illustrations may seen somewhat oversimplified, they are extremely common occurrences.

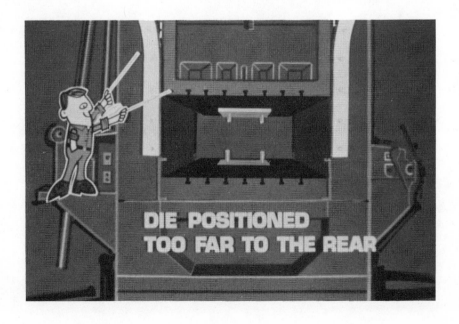

Figure 13

In Figure 14, the die is too small and positioned too far to the right.
The deflection of the machine shown here is excessive either because of
improper location or size of the die, or both.

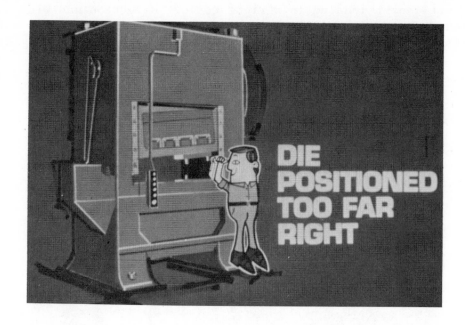

Figure 14

Figure 15 shows an actual photo of a die that is too small for the press
and located too far to the right. This is a full press load under only
one connection.

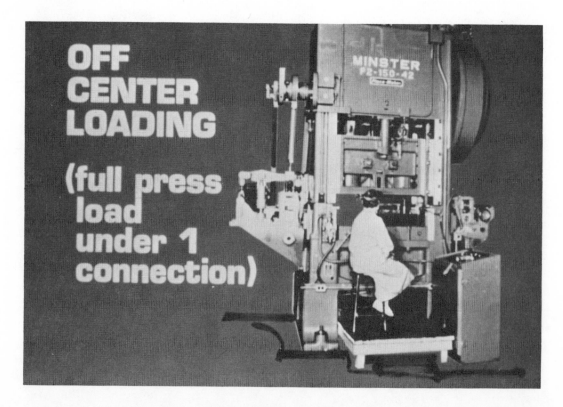

Figure 15

Figure 16 illustrates a small die in a large press. The most frequent mis-application usually occurs when the small die requires a high tonnage and the only high tonnage press available is one with an excessively large bed area. The net result is disaster.

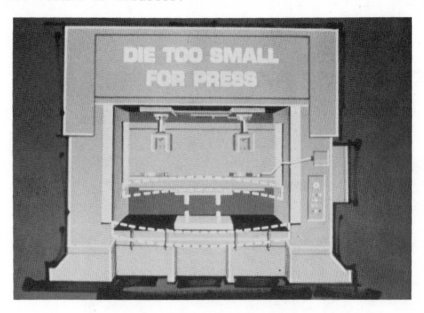

Figure 16

Figure 17 illustrates the way a die should fill out the machine. In other words, the die and the machine should be properly matched.

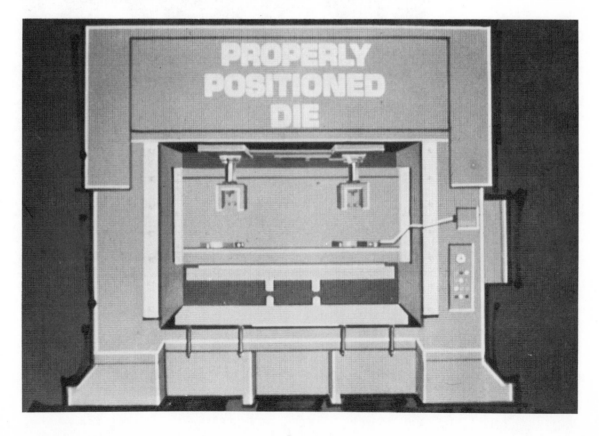

Figure 17

The "C" frame press deserves special attention because many progressive dies are placed in this type of equipment. Again, the die can be positioned too far to the rear, or too far forward as shown in Figure 18. In the "C" frame design, proper location of the die is much more important and more critical than in the straight side machine.

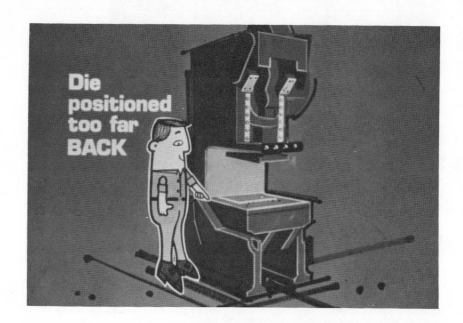

Figure 18

Here is what happens. "C" frame, single point machines are not designed for any type of off-center loads. The only restraining influence that holds the die in a parallel position is the frame and the gib. Neither of these members are capable of restraining the die completely, and, in such a case, more strain is placed on the press than it is capable of handling, as shown in Figure 19.

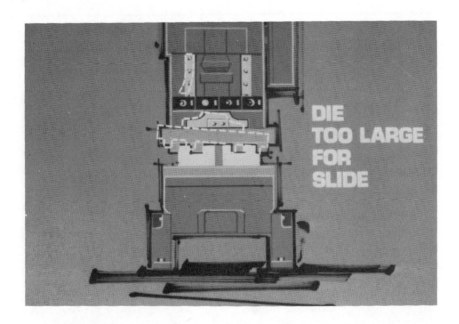

Figure 19

A "C" frame machine deflects a great deal more than a straight side press. For example, the normal 100 ton "C" frame machine will deflect approximately twenty-five thousandths of an inch (0.65 mm) for a distance measured from the edge of the gap of the "C" frame to the center line of the slide. Any overloading beyond the rated capacity causes severe deflection. It is not unusual to see these overloaded machines running with one-sixteenth of an inch deflection. Poor quality parts are produced because of such conditions. The frame actually deflects as shown in Figure 20.

Figure 20

There are two basic types of press drives that require a thorough understanding. The slide in Figure 21 illustrates the most common type, which is referred to as the flywheel drive. In other words, the drive is direct from the motor to the flywheel and the flywheel turns directly on the crankshaft. This type of drive is commonly used for work performed near the bottom of the stroke, such as blanking, piercing and light forming operations. The energy available -- not the tonnage -- from such a drive is small. The tonnage in such a drive is the rated capacity of the machine. For example, 100 tons. But the rating of the 100 ton flywheel machine is near the bottom of the stroke, never greater than 1/8" (3 mm) above the bottom of the stroke.

Figure 21

Figure 22 illustrates an ideal and extremely common application of the flywheel type drive. The flywheel type drive lends itself to high speed operations and is ideal for supplying the necessary tonnage to produce the rotor and stator stamping.

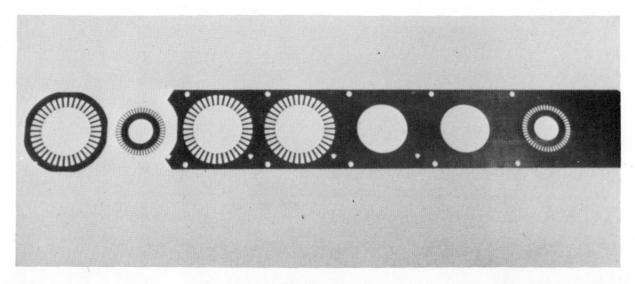

Figure 22

For heavy thickness material, or drawing work, it is necessary to reduce the speed of the machine. However, the flywheel of the machine cannot be reduced in speed because this would reduce the energy level. The press to choose, then, is a geared type drive which allows the ram speed to be reduced to the proper velocity, at the same time allowing the flywheel speed to be maintained through the gear ratio. (Figure 23). The net result is a high energy machine. Compared with the flywheel type drive, the geared drive would have a rating quite a bit higher in the press stroke. In fact, most gear drive presses are rated at one-quarter inch (6.3 mm) above the bottom of the stroke. The geared machine, at points greater than a quarter inch above the bottom of the stroke, say one or two inches (25.4 or 50.8 mm), will have a reduced energy level. But, even though they do have a reduced energy level, their capabilities at this point in the stroke are at least ten times greater than the flywheel type drive at the same point above the bottom of the stroke.

Figure 23

A perfect example of where a geared machine must be applied is shown in Figure 24. The part produced by this progressive die has a drawing operation approximately an inch and a half (38 mm) deep from materials sixty thousandths of an inch in thickness. The tonnage of this job might be approximately two hundred or more tons, but that is not important. What is really important is the energy level. It is vital that the engineer who has designed this die, select a press with the proper tonnage and the proper energy level so that the work can be performed. The part in this illustration ran at a speed of approximately sixty strokes per minute on a three hundred ton press with a ten inch stroke. The ten inch stroke was necessary in order to have enough stroke to feed the part using a double roll feed. The combination of these specifications then dictated the proper energy level to design for the machine.

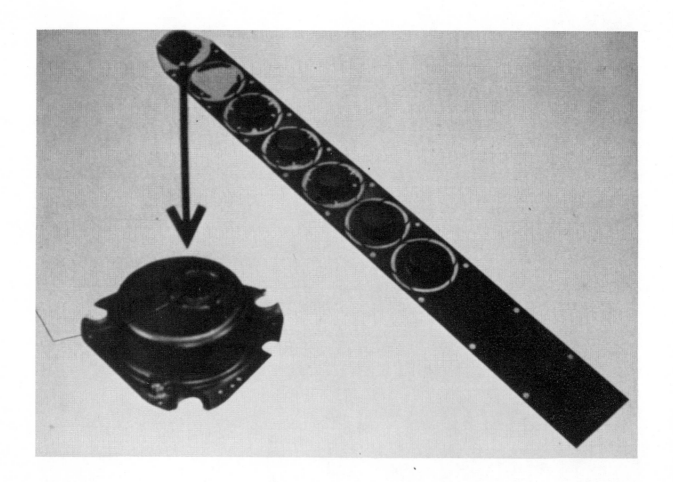

Figure 24

The diagram in Figure 25 shows the relation of the capacity of the press at various points in the press stroke. The further the slide is from the bottom of stroke the lower the capacity of the press.

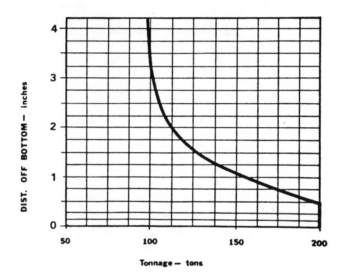

STROKE-TONNAGE CHART

200 TON

8 inch   STROKE

RATED $\frac{1}{2}$" OFF BOTTOM

Figure 25

The V-gib shown in Figure 26 is typical of the type of gibbing required for the "C" frame design machine. This type of gib will only support the slide when the load is applied in the center of the slide. This design is not recommended for off-center loading which is commonly encountered in progressive dies.

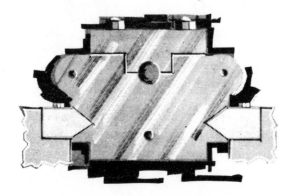

"V" GIB

Figure 26

Figure 27 shows the six point gib, which at the front area closes in the gibbing with a box type arrangement, and at the rear area uses a 45° gib. This type of gibbing is commonly employed on the standard straight side press, however, there are better arrangements than this for progressive die applications.

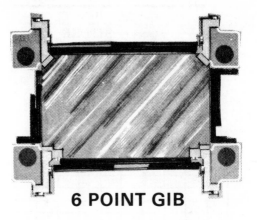

6 POINT GIB

Figure 27

The eight point gib shown in Figure 28 is the best that can be applied for progressive dies because this gib tends to resist any rotating movement of the slide caused from off-center loading that is encountered by most progressive dies.

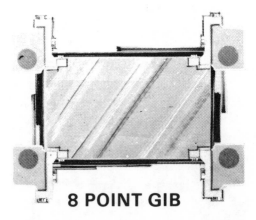

8 POINT GIB

Figure 28

Figure 29 displays the three most common types of crank drives used with mechanical type presses. The illustration on the left is the standard crankshaft, in this case the crank used is a double or a two point crank design. The illustration in the center is the eccentric gear drive which does not employ a crankshaft in the customary manner. The motion of the slide is obtained by large eccentrics that are an integral part of the drive gear, hence the name eccentric gear drive. This design is used with the larger long stroke type of presses. The illustration on the right is a modification of the standard crank drive which uses a full eccentric for the throw rather than the off-set crank pin. This drive is best used with progressive dies because the elimination of the crank pin in favor of the full eccentric more fully supports the crank and prevents deflection in the crown of the machine.

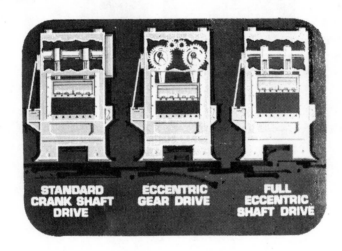

Figure 29

In Figure 30, the illustration on the left shows the standard crank design and may be compared with the illustration in Figure 29. The illustration in the center shows the eccentric gear design; notice that the eccentric is cast as an integral part of the gear and that this design replaces the crank throw seen on the left. The illustration on the right is the modified crank design which employs the full eccentric. It is now possible to bring the support of the bearings right up to the connection so that the unsupported area no longer exists and minimum deflection characteristics are achieved.

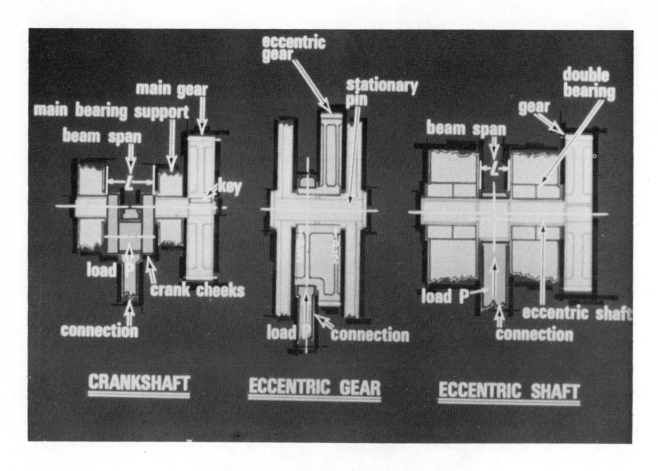

Figure 30

In Figure 31, the illustration on the top shows the common bearing arrangement for the standard crankshaft design. In contrast, the illustration at
the bottom shows the sturdy support achieved from having the bearings right
up to the connection on the eccentric shaft design.

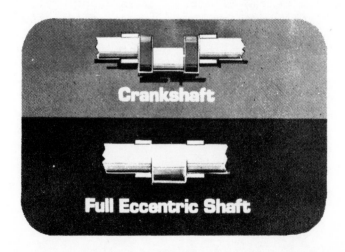

Figure 31

The double crank full eccentric shaft (Figure 32), shows the complete bearing arrangement and the support which it gives in detail.

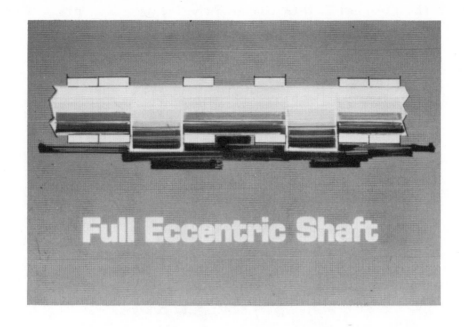

Figure 32

A press must have energy, i.e., the ability to perform work in order to produce formed and drawn parts. To correctly apply a press, therefore, the knowledge of available energy in a press is necessary to determine if the machine has enough force to do the work. The energy required from a press is simply the product of the required force (load) times the distance the force must be exerted as illustrated in Figure 33.

PRESS ENERGY CONCEPTS

To make a part we need

FORCE (Tonnage) thru a DISTANCE (Inches)

F x D = WORK

ENERGY = Capability of a body to do WORK

So where do we get ENERGY in a press drive ?

Figure 33

Of the available kinetic energy in a flywheel only a portion of this energy may be used. Total energy in a flywheel is the product of the flywheel mass times the flywheel velocity. The energy removed from the flywheel may be simply calculated after first determining the change in velocity of the flywheel. This change in velocity takes place when the flywheel gives up its energy to perform work. (Figure 34).

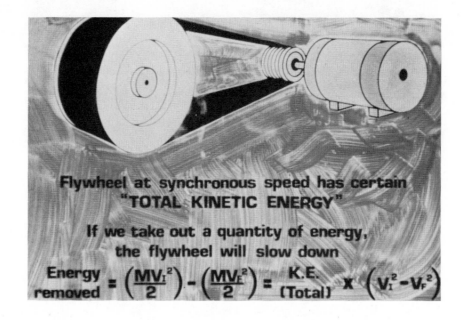

Flywheel at synchronous speed has certain "TOTAL KINETIC ENERGY"

If we take out a quantity of energy, the flywheel will slow down

$$\text{Energy removed} = \left(\frac{MV_I^2}{2}\right) - \left(\frac{MV_F^2}{2}\right) = \frac{K.E.}{(Total)} \times \left(V_I^2 - V_F^2\right)$$

Figure 34

A common straight side press has its flywheel located as shown in Figure 35. The flywheel is in the center of the photograph driven direct from the electric motor by a V-belt drive.

Figure 35

The electric motor must restore the energy removed from the flywheel. The motor restores energy during the non-working portion of the press stroke. The time available for restoration begins on the upstroke of the press and ends at approximately the midpoint of the downstroke. (Figure 36).

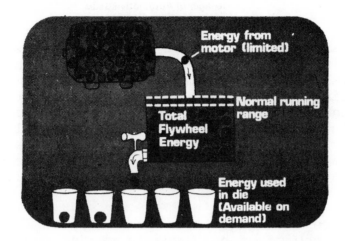

Figure 36

The flywheel cannot contribute energy at a rate greater than the time available for the motor to restore the energy. Figure 37 illustrates what happens if the demand on the flywheel exceeds the motor output. After a few strokes of the press the flywheel speed is reduced to a low energy level and the motor speed is below its synchronous range. The result is the motor becomes very hot and the press no longer is capable of doing work.

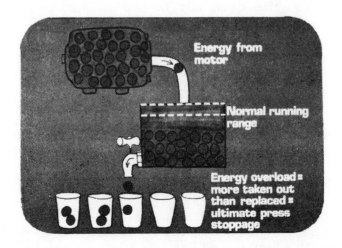

Figure 37

Presented at the Transfer Die Technology Workshop, February, 1970

Selecting A Press For The Transfer Die

By Philip V. Monnin
Manager of Automotive Sales
Minster Machine Company

There are two basic "ways to go" in terms of transfer techniques in metalforming presses -- one is a special press with integral transfer mechanism and also "somewhat" integrated tooling. The second is a reasonably standard press with a "transfer die." Perhaps we should define "transfer die" in this case as transfer tooling with self-contained transfer devices, which can be installed or removed from the press as a basic unit. This "transfer die" is synchronized to the press motions with simple mechanical arrangements.

There are advantages and disadvantages in both techniques. The choice of one type in preference to the other depends on the resultant evaluation of many factors; among them -- number of parts to be run, the production speed required, length of part runs, total expected production, cost of the equipment required, familiarity of user's people with techniques involved, user's maintenance capability, basic types of parts being run, ultimate press usage after the high volume jobs are gone, etc. Obviously, this list of factors could be continued into more detailed discussions and great controversy could erupt on which is the better approach. It is not the purpose of this paper to become argumentative but to constructively investigate application criteria for the standard press -- modular transfer die concept.

The name of the game in today's business operation is versatility!! When investigating the use of transfer concepts, this one word should guide basic decisions. This paper will be basically oriented in developing press specifications that will insure adequate press capacity and maximum press "versatility!"

In this paper the following topics will be covered: type of press, basic press specifications, tonnage, energy, sizing a press and press options for maximum versatility.

TYPE OF PRESS:

It is sometimes difficult on the part of the user to choose the type of press really needed on a particular job. He, many times, simply relies on the press builder or on a

salesman who "happens to have one available." Several factors
must be considered before choosing the type of press--but first,
it is necessary to establish a guideline of evaluation. The
optimum situation, in any case, is maximum die life, maximum
production speed, the ultimate in part accuracy and minimum
down time for tool or press repair. To achieve all these items
is "utopia," and obviously the most desirable situation to be
in.

How can a press choice affect the above points? What
press qualities does the die want to see?--Basically, the press
must have inherent parallelism and rigidity to withstand off
center loading. The punches must enter into the die cavity
with contact surfaces as parallel as possible. An OBI or gap
press cannot possibly provide as good parallelism and off center load resistance as a straight side press.

As indicated on Figure 1, a two point machine has more
inherent rigidity/parallelism than a single point press.

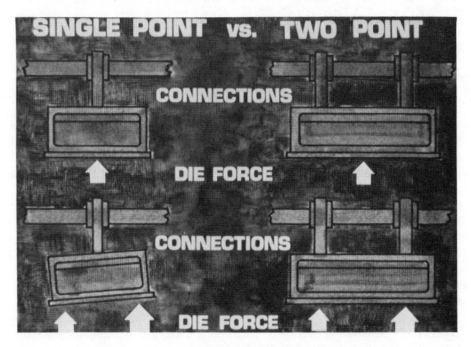

Figure 1.

We also recommend that twin drive be specified with special
opposed helical main gears. The twin drive provides additional
tonnage capacity off bottom of the stroke and better parallelism by reducing torsional windup of the main shaft. The
opposed helican main gears, when combined with a specific shaft
float, actually transfer a portion of the off center loads to
the opposite end of the eccentric shaft away from the heavily
loaded end. The end result of the above is better parallelism
under off center loaded conditions--a critical consideration
in transfer applications.

To the above points we would add 8 point gibbing, and recirculating oil lubrication system. These features contribute to minimum clearances and maximum accuracy.

The basic press type we have described above is commonly referred to as a "progressive die press" by most press manufacturers and generally represents the most accurate presses in their product line; generally it will be a two point, straight side machine. For example, slide to bed parallelisms of .004 T.I.R. maximum or less are obtainable for presses with beds of 84 x 48 and tonnages up to 800 Tons. Remember that the "progressive die" label does not limit the press application--a transfer die is really a progressive die, or you may say a progressive die is really a transfer die--it is relative to the side of the fence you are on.

Several drive types are available: Crankshaft, eccentric shaft, and eccentric geared, to name a few. Each has its advantages and some have distinct disadvantages. Again, this is an argumentative subject and beyond the scope of this paper. The best advice is to evaluate the drive on an engineering basis and investigate parallelism, the press's ability to withstand off center loading, and the machine clearances.

Bed and slide deflection is critical for die life and should be held to a minimum. Maximum recommended bed deflection is .0015 inch per foot of die space with full tonnage evenly distributed over 2/3 the right to left die space. The slide deflection should not exceed .001"/ft.

All of the above comments are general and not specifically required in every application but should be considered if die life, part accuracy, maximum speed, and minimum die life are important in your operations.

BED SIZE:

One of the basic questions to be answered is "how large of a press is needed for this job?"

Figure 2 shows several items that must be considered in choosing the press bed (also bolster and ram) size.

First of all, what is the die size you plan to run? What will future die sizes be? Will the transfer mechanism or drive require additional space? How are you going to feed material to the press--blanks?--coil? Will this require additional bed space? How are you going to get the parts out of the die? Will this require bed area?

All of these questions must be answered before establishing bed area. Notice that on Figure 2 there is also a block called "GOOD-T" space. What does it mean?--an important consideration--"Get Out of Deep Trouble" space. This is the answer to your prayers if you have an unanticipated problem in tooling and need additional space. Sometimes the "GOOD-T" space is the most economical portion of the press and also the most costly when you haven't considered it prior to purchasing the press equipment.

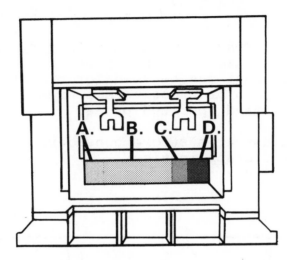

A. **Mechanical transfer or feed**
B. **Basic die requirements**
C. **Part removal**
D. **"GOOD-T" space**

Figure 2.

SHUTHEIGHT:

In the above section, bed (slide and bolster) area factors were considered. We must now determine the vertical die space required. This "vertical space" is commonly referred to as "shutheight" and is measured in inches slide to bed or slide to bolster S.D.A.U., (stroke down-adjustment up),--be sure to specify either "to bed" or "to bolster."

Another factor to be considered along with the required shutheight is the amount and type of slide (ram) adjustment. Transfer applications can require more shutheight adjustment than other dies, especially since "versatility" is a prerequisite. Slide adjustment ranges from only a couple of inches to 6", 8", 10", 12", 16" and, in some special cases, more. If "non-guided" and "non-supported" adjustment systems are used by the press builder, the adjustment should be kept to an absolute minimum to reduce the chance of screw damage, and excessive "out of parallelism" under load. We recommend a "guided barrel" type connection and screw assembly--it eliminates a major component of bending load in the screw when work is contacted "off bottom" of stroke. The end result to the user is considerably better accuracy and greatly reduced maintenance in the slide adjustment system. The "guided barrel" arrangement also permits longer adjustments while still providing the above advantages. At the same time, we still recommend that the adjustment be kept as short as possible--typically 10" to 12" is a good "maximum" choice.

STROKE

In terms of stroke, versatility is very important. Since a "modular transfer die" can easily be removed, it is advisable to choose a press that can be used for transfer and also can be easily converted to a progressive die press or even to a "single stroked", hand operated press.

Figure 3 illustrates stroke considerations for the typical modular transfer die application.

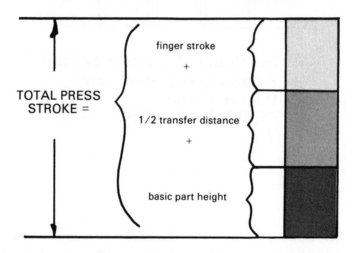

PRESS STROKE—TRANSFER DIE

Figure 3.

The following rule of thumb applies to most applications where transfers are press actuated:

Press Stroke = transfer finger stroke + 1/2 transfer
                                          distance
                                  + 1/2 basic part
                                          height

The Stroke for a progressive die application is illustrated on Figure 4.

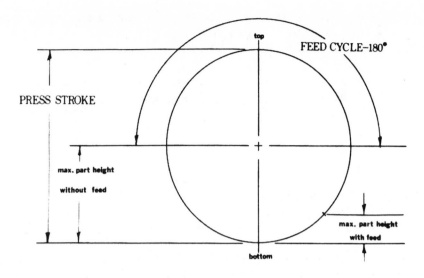

PRESS STROKE = 4 times part height

PRESS STROKE - PROGRESSIVE DIE

Figure 4.

Note that the "safe" rule of thumb is: Stroke = 4 times part height, when using a feed with a 180° feed cycle. A shorter feed cycle would enable a shorter press stroke. To get the part out of the die, a stroke of at least 2 times part height is required with manual loading (no feeding mechanism).

After considering both the transfer and progressive applications (both present and anticipated), choose the longer of the two strokes for maximum equipment versatility.

SPEED:

Speed is a difficult item to establish because it is dependent on many variables.

Generally press speed does not limit the speed of operation. At least to date, press speeds beyond die and transfer capability are readily available. Typical single geared-twin drive press speeds are obtainable as follows:

400 Ton with 10" Stroke-- 90 S.P.M.
400 Ton with 7" Stroke--120 S.P.M.
200 Ton with 5" Stroke--150 S.P.M.

The main limiting factor for operational speed is die capability and transfer design. Exceeding the critical speed of draw can have disturbing results.

This same problem can be the result of improper die design or inadequate lubrication. Trying to transfer the part too far, too fast can also be a problem.

In selecting a press speed range, it is best to investigate the transfer and die speed capability very closely. Speed capability should be a prominent criteria for choice of transfer design and the approach to actual die design.

TONNAGE:

A press can be overloaded in two distinct and separate ways: excessive tonnage and excessive energy usage. It is possible to have one type of overload and not the other.

The tonnage rating of a press is a relative measure of: 1) torque capacity of the drive; and 2) structural capacity of the frame. Tonnage is rated at a certain distance off bottom and is represented graphically as shown in a typical Tonnage vs. Distance off bottom of stroke curve shown on Figure 5.

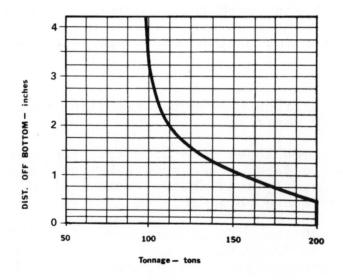

STROKE-TONNAGE CHART

200 TON

8 inch   STROKE

RATED ½" OFF BOTTOM

Figure 5.

A common misconception is that a 200 Ton mechanical press will provide 200 Tons at any position of stroke--this is wrong and dangerous. Due to basic crank and slide relationships, the tonnage capacity drops off rapidly at points above the bottom of the stroke. This is true for all mechanical presses regardless of manufacturer. Note that Figure 5 illustrates a tonnage of 125 Tons capacity of 2" off bottom of stroke. In other words, a die requiring 130 Tons @ 2" off bottom would overload the clutch and drive mechanism of this particular press. Also note that this press will provide 200 Tons up to 1/2" off bottom and would be rated as 200 Tons @ 1/2". Other standard ratings are 1/4", 1/8", 1/16" or less, depending upon press type and manufacturer's design criteria. Ratings over 1/2" are special cases and not generally provided except on special request.

Again referring to Figure 5, any tonnage vs. distance off bottom point to the left of the curve would be within tonnage capacity--any combination to the right would be a tonnage overload for the drive.

The structural portions of the press are designed for specific deflections at rated tonnage and are designed with various safety factors, depending upon the press manufacturer's design criteria.

ENERGY:

What is energy in a press? Why is it needed? Where does it come from? These are questions to be answered.

In a normal drawing die, the punch contacts work at a certain point above the bottom of the stroke. At this point, a certain tonnage is required to start the drawing process. This tonnage working thru a distance results in a need for energy to finish the part operation. We must have a source of energy in the drive of the press. That is the reason the press is equipped with a rotating flywheel.

The following is one of many cycles in the life of a press flywheel:

1.) The press is started and the motor rapidly accelerates to its synchronous speed (Remember that the motor only knows one thing--stay at synchronous speed).
2.) The clutch is actuated and the press cycle starts. Toward the bottom of the stroke, the part is hit and the draw begins.
3.) The draw requires energy which is taken from the Kinetic energy of the flywheel as shown on Figure 6.
4.) The flywheel slows down in the relationship shown on Figure 6.

5.) The motor slows down at the same time but immediately starts to accelerate trying to get back to synchronous speed.

6.) It makes it back up to speed and the press is passing the top of the stroke ready for the next hit.

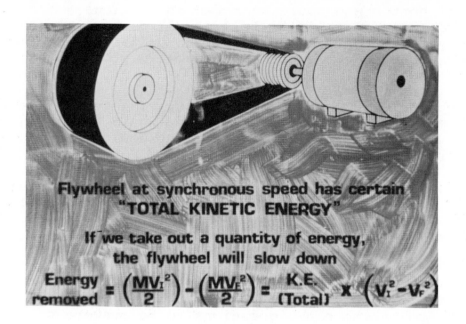

Flywheel at synchronous speed has certain "TOTAL KINETIC ENERGY"

If we take out a quantity of energy, the flywheel will slow down

$$\text{Energy removed} = \left(\frac{MV_I^2}{2}\right) - \left(\frac{MV_F^2}{2}\right) = \text{K.E. (Total)} \times \left(V_I^2 - V_F^2\right)$$

Figure 6.

Figure 7 shows this procedure in schematic format. The amount of energy removed from the flywheel is generally only a small portion of its total kinetic energy.

Energy Overload is depicted in Figure 8, which illustrates what happens when more energy is removed from the flywheel than the motor is capable of replacing each stroke of the press. The ultimate result is press stoppage and potential motor damage.

In many cases, a transfer press will require more energy available to do work than is available in a standard drive. It is essential to advise the press builder of the application so that additional energy capacity can be added to the drive.

Another rule of thumb is that a standard press will not have more capacity than its tonnage capacity times the rating off bottom. For example, a 200 Ton press @ 1/2" off bottom would not be capable of more than 100 inch-tons energy available to do work on a continuous basis. In fact, the energy capacity will usually be slightly less. If the energy calculations for a particular job approach or exceed this "rule of thumb" rating further investigation and precautions are advised.

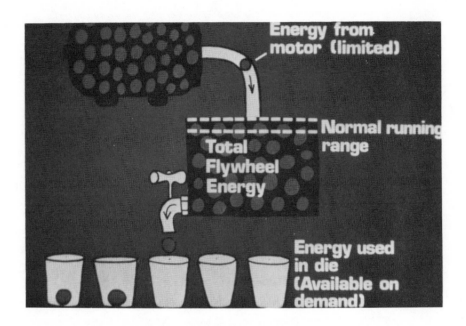

Figure 7.

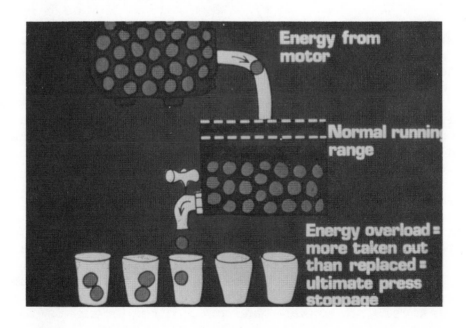

Figure 8.

A common question by many press users is what does "full energy point" mean? Basically, this concept applies to presses with variable speed drives. Full energy point is the lowest press speed (strokes per minute) at which the rated energy of the press can be removed without causing an "energy overload" condition as described above. This speed-energy relationship is shown on Figure 9.

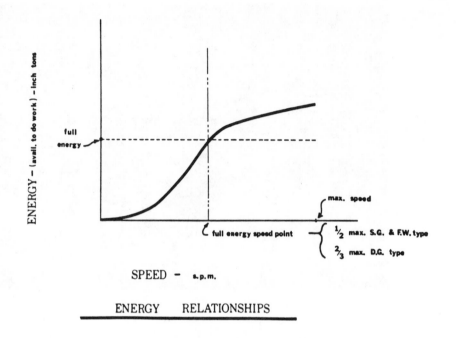

ENERGY   RELATIONSHIPS

Figure 9.

For example, if we have a single geared press running from 0-60 SPM, the full energy point will be at 30 SPM. In the case of a double geared press running from 0-30 SPM, the full energy point would be 20 SPM.

Variable speed drives are very beneficial in that optimum speeds can be determined for each individual die/transfer set-up. The end result is maximum production and maximum machine utilization.

DETERMINING PRESS CAPACITY.

Determining press capacity for a transfer or progressive die can appear to be difficult at first glance--primarily because so many operations are occurring at one time. A way to simplify analysis is to take each station individually. On Figure 10 (next page), a simple chart form is provided to enable analysis of tonnage and energy requirements for multi-station dies. Sample calculations will be performed on a multi-station transfer for an automotive suspension part shown in Figure 11.

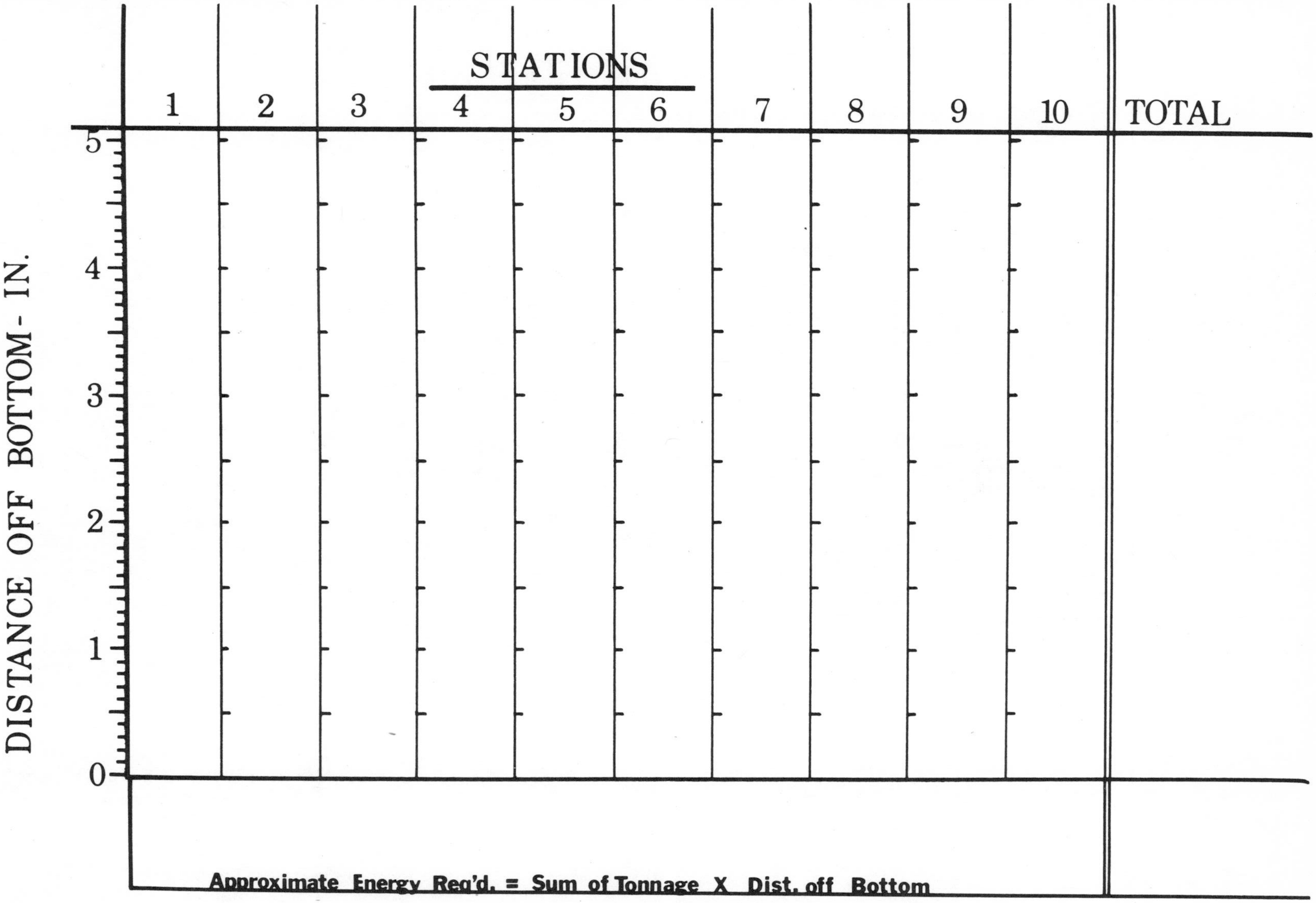
STATIONS
1
2
3
4
5
6
7
8
9
10
TOTAL
DISTANCE OFF BOTTOM - IN.
5
4
3
2
1
0
Approximate Energy Req'd. = Sum of Tonnage X Dist. off Bottom

PRESS CAPACITY CHART

Figure 10.

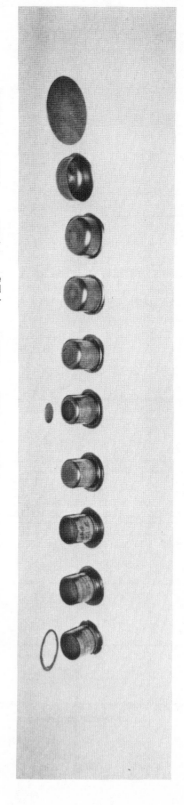

PRE-
BLANK

STA.
1

STA.
2

STA.
3

STA.
4

STA.
6

STA.
5

STA.
7

STA.
8

STA.
9

NOTE: Items pictured in stations
5 & 6 have been reversed.

Figure 11.

In this particular die, a blank is fed into station one and a series of multiple operations are performed in stations 1 thru 9. The resulting analysis is shown on Figure 12. Note that station 1 is a draw requiring 18.8 Tons at 1" off bottom. The energy required is 18.8 inch-tons. Also note that station 7 has three operations which are recorded separately.

When all stations have been analyzed, the tonnages are added up across the page into the right hand column of the chart. The approximate energy requirement is the sum of all the individual energies along the bottom of the chart.

In the final press sizing for this particular part, note that all the total tonnages fall to the left of the tonnage capacity curve of Figure 5. The maximum tonnage requirement is for 154.2 Tons near the bottom of the stroke. The total energy requirement of 113.38 inch-tons indicates that a 200 Ton @ 1/2" standard press would not have adequate available energy and that a high energy drive would be required.

The above is a good example of how an improper press choice can be made if energy is not considered.

INCREASING PRESS VERSATILITY:

The following is an outline of options and points to consider when determining press specifications. These items will improve versatility in many applications:

1.) Feeding Arrangements: Eliminate separate blanking by feeding right to left or front to back into die. Consider future use of press with progressive dies. Consider skip (traversing) feeder for material savings.
2.) Press Controls: Available on the press upright, in a pedestal off press, or a movable pendant on press. Relay or solid state available.
3.) Die Safety Controls: Several on market, a necessity for good die protection and maximum speed.
4.) Overload Detection: Motor load meter to indicate relative energy usage. Tonnage indicator to indicate tonnage being developed in dies.
5.) Jamming Devices: Hydraulic Overloads not recommended because of ill effects on parallelism, maintenance and speed limitations. Hydraulic tie rod nuts recommended to enable fast unshrinking and reshrinking of tie rods.
6.) Transfer Drives: Low cost options to provide 1:1 drive for exact transfer timing and smooth operation.
7.) Cushions-lower: Available fixed or sliding type.
8.) Cushions-upper: Available as built-into-slide or built into a removable slide sub-plate.
9.) Upper Knockouts: Available as positive cam actuated type or air loaded for predetermined pickup point.

DISTANCE OFF BOTTOM - IN.

PRESS CAPACITY CHART

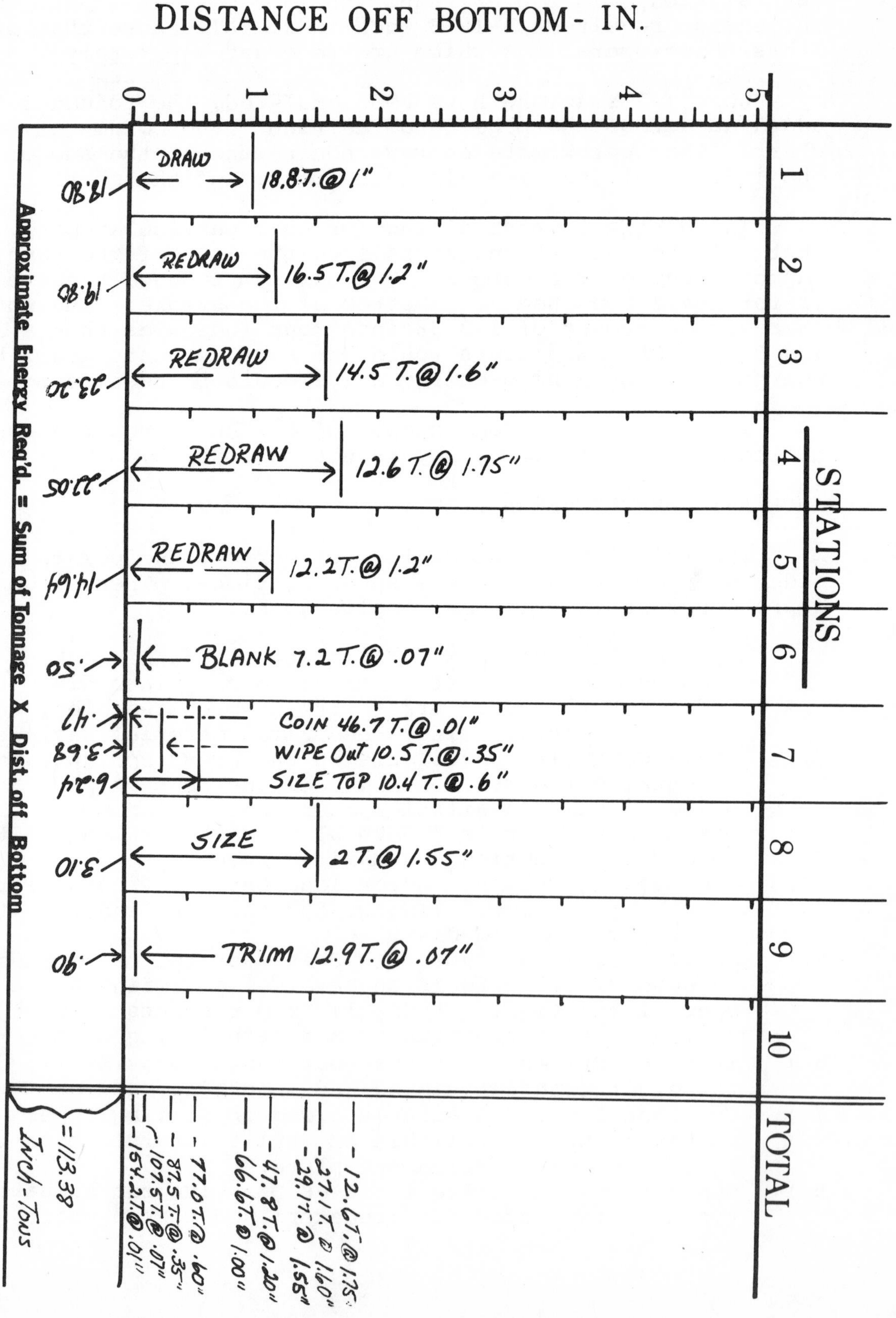

Figure 12.

62

10.) Lower Liftout: Available as a rotating type or as a reciprocating type.

11.) Die-Flote: Low cost air flotation methods for easy location and/or removal of dies or bolster assembly from press.

CONCLUSION:

Hopefully this paper indicates that careful consideration must be given to press choice for transfer and other applications.

It is very easy to choose the wrong type, size, tonnage, energy capacity, and options for a press to be used on transfer applications.

With proper application, total operation performance can be considerably better than the "use what's available" approach to press choice.

To enable proper application, it is important to bring the press application people into the picture as early as possible and to provide them with all the application data at your disposal. This cooperation will certainly help your equipment "versatility" and "utilization."

Total System, Heavy Tonnage, High Production Transfer Presses

By Walter C. Johnson
Vice President, Systems Marketing
Verson Allsteel Press Company

To fully understand and appreciate the extent of this type of total system, we will identify each of the elements for the system.

Heavy Tonnage - A press size range from 600 tons to 6,000 tons. The elements of design and the limitations to size are important. Structurally, size cannot go beyond what can be manufactured or shipped or erected at the plant site. Machines of this size range were not produced fifty years ago as the techniques of metal fabricating, welding, and machining had not been developed to even visualize such possibilities. From an investment point of view, a 600-ton press is in the million dollar range and a 6,000-ton or larger machine could be five or more million, plus the foundation, installation and ancillary facilities. Presses in this size range have been built to produce, automatically, stampings up to forty inches square. They have been built to produce from coil or blanks of light gauge aluminum to 5/16" steel coil stock and from steel blanks up to 3/8" thick.

High Production - is relative to the size of press and shape of the stamping, complexity of the operations and the ability of the material to withstand the forming or drawing or cutting. A small stamping weighing a pound or two can be transported more rapidly than a stamping weighing twenty to thirty pounds. Also, the small stamping is usually produced in a smaller size machine where the press elements are smaller and the inertia of the moving elements is less. A heavy stamping in the larger press is more difficult to control as are the press components.

The transfer press in itself is a metal forming machine that permits multiple operations on a stamping which could not be completed in one die and thus will perform secondary operations automatically in the one machine. A transfer press system requires a minimum of floor space, reduced power consumption, a minimum of labor with a resultant production rate beyond the normal capability of a manual operation.

The total system is a combination of right size press,

proper tooling design and the elements of transfer press features which go into an automated production system.

The automotive industry and the major appliance manufacturers employ most of the large tonnage transfer presses. These two industries, including millions of cars and trucks and millions of refrigerators, clothes washers, dryers, dish washers, etc., require production that dictates the absolute requirement for the most productive and sophisticated machine tools available. Today's transfer press answers that need.

The determination to consider and select a transfer press requires an entirely different approach than the purchase of a conventional metal forming press. To produce a specific stamping, there are three conventional approaches. First, the progressive die method which calls for a coil-fed press and a progressive die. This is the least expensive approach as long as the stamping can be carried in the coil strip to completion and if the material scrap loss is not prohibitive, the second system would be a series of presses with press-to-press automation, this may be the logical approach if the stamping is very large such as an automobile roof or a hood panel. However, productive speed is limited due to the many mechanical events which must occur with each individual press stroke. The third is the transfer press. A transfer press reduces the scrap loss of a progressive die setup and permits drawing which cannot be accomplished to any great depth in progressive tooling. The transfer press also eliminates the individual press line and press-to-press feeding, whether it be hand-fed or automated.

Typically, one of the largest producers of refrigerator and air conditioning compressors recognized that to maintain its competitive position in their market it would be necessary to increase production facilities and reduce labor input. Instead of adding a line of new presses, a transfer press was considered. A compressor housing stamping can require eight to ten operations from blank to finished part. One transfer press combined all operations and the press is operated by one operator.

The selection of a transfer press requires the very careful analysis of the piece part or the family of parts under consideration. The study must start with a blank or coil selection. The part must come from a coil or strip with the absolute minimum of scrap waste. If this is not possible, then consider using pre-cut blanks which have been made by "nesting" or stagger blanking. Each individual operation must now be determined to complete the part and with each operation the individual tonnage calculated along with die cushion tonnage or stripping tonnages. The depth of the part will determine the press stroke and the speed at which the machine can operate. Finally, the size of each die will establish the overall size of the press die space based on the number of operations required.

A simple layout of this data will illustrate the overall press loading. Overlaying the same information of other stampings considered for the press will further determine the machine specifications for the range of stampings. We cannot overemphasize the importance and value of this data when making a machine selection. In a transfer press, machine loading is all-important.

The ideal condition would be to arrange the tooling in the press so that the individual tonnage would be evenly distributed over the total die area. However, this is rarely possible. Therefore, the press size and design must be keyed to suit the conditions.

As an example, if the total tonnage to produce a stamping is 2,000 tons and 1,500 of that total is in the first two die stations with the remaining operations requiring 500 tons, a 2,000-ton press would not be practical. A 3,000-ton press would be necessary to prevent serious off center loading at the heavy load side. In the single slide press an off center load condition would still exist and, while not necessarily detrimental to the machine, will cause a tilting action of the slide due to normal bearing, gearing and gibway clearances. This tilting action, however small, could have a detrimental affect on the quality of a precision stamping.

When tooling cannot be rearranged to balance the overall load, it might be more practical to consider a three-column design press. In effect, what happens is that the press is divided into two parts: in this case of our example, a 1,500-ton section and a 500-ton section with a center separating column. The press would still have a single one-piece bed and crown but individual slides driven by the common drive train within the crown. The press tonnage would be less and we have eliminated the out-of-balance load condition.

The design of the press frame is important to review again and to be more specific, our reference to multiple column machines. Normally, a metal forming press is made up of a bed or base which supports the press and the tooling, plus two side columns or uprights to support the crown of the press. Between the uprights is the slide or ram. The bed, columns and crown are held together by the tie rods.

Illustration No. 1 is of such a press. It is a 1,000-ton capacity press with a total die space area of 5 feet front to back and 20 feet right to left. There are six die stations on forty-inch centers. The press produces stampings for an appliance manufacturer at a rate of 1,800 hundred parts per hour.

Illustration No. 2 is a 3,500-ton single slide press with eleven die stations, each on twenty-four-inch centers and a die area of 5' by 23'. The press automatically produces 1,200 automobile torque converter housings per hour.

ILLUSTRATION NO. 1

ILLUSTRATION NO. 2

To produce stampings with extremely unbalanced load conditions, it is necessary to design the press in a manner which will provide the best load conditions possible for the benefit of the machine life and total performance of stamping quality.

Illustration No. 3 is a three-column 3,000-ton press. The stamping is the lower control arm of which two are used in each automobile. There are twelve of these presses in one line producing this automobile part. The press is divided into two die space areas. The right-hand slide is 2,000-tons capacity with six die stations on thirty-inch centers. The left-hand slide is 1,000-tons and has seven die stations. The major tonnage load is in the first four operations. If this press had been designed as a single-slide press, the machine would be at least 4,000-tons capacity with a single ram over thirty-eight feet long. It is doubtful whether such a machine could be shipped.

Illustration No. 4 is another example of a three-column machine. It produces a variety of precision stampings which go into the automobile transmission. It has twelve die stations: six at the 2,500-ton right-hand slide and six at the 1,500-ton slide. It produces one thousand to fifteen hundred complete stampings per hour. The press stands thirty-feet above the floor and weighs over 1,800,000 pounds.

The design of a press with more than two slides and three columns would be the exception rather than the rule; however, automotive production requirements can be that exception and justify the need. An example would be illustration No. 5. This is a 4,600-ton transfer press with four columns. The right-hand slide is 2,000 tons and has seven die stations. This illustration is a rear view of the machine so the right-hand slide is on the left of the photograph. The center slide is a single die station of 2,000 tons and the left slide is 600 tons with four die stations. The stampings produced are automotive brake backing plates which require a heavy coining in the center slide. In this particular die area a hydraulic cushion of 2,000-ton capacity provides the necessary overload protection to the press and the tooling. In each of these illustrations it is important to note that the press is driven by a single motor and a single flywheel which transmits the required energy to the gearing within the crown. The single energy source insures that all slides are driven simultaneously at the same speed. Regardless of whether it is a two-point single slide press or a five or six-point three-slide press, all driving gears are interlocked and operate as a single unit.

Stock feeding of a transfer press for metal forming work is either by a magazine or stack feed which feeds precut blanks or by a coil feed from coils into the first die station. When feeding blanks, stacks of blanks are moved into position either by a dial magazine or in-line on a roller conveyer. The stack is then elevated to feeding position by a lower cylinder. The

ILLUSTRATION NO. 3

ILLUSTRATION NO. 4

blanks are then picked off from the top by either suction cups or a magnet. The feed fingers of the transfer then carry the blank into the first die station.

Coil feeding of transfer presses is usually recommended when the blank is square or rectangular and scrap loss is minimal. If the blanks are round or irregular-shaped, it could be wise to consider a joggle or zig-zag type coil feed. For round blanks, it is possible to realize a material savings of over 6% or a reduction in scrap of 31%. On some shapes, the material savings can go as high as 12%. Furthermore, it now permits using double width coils and reduces downtime for coil changing by fifty-percent.

Illustration No. 6 shows the coil strip and the skeleton of how the strip is moved front to back over the blanking die station.

Illustration No. 7 is of a smaller 450-ton press with an unusual combination of material feeding options. On the extreme left-hand end of the press is a three-station dial-type stack feed for precut blanks; along side the outside of the left-hand press column is a front-to-back coil feed. As part of the press design, an outside blanking station was provided driven by the main press drive. Mounted to the outside blanking slide is a joggle die unit which permits zig-zag blanking as an option. Consequently, with this press it is possible to feed either blanks or coil stock.

The transfer feed itself is the major key to efficient, reliable and trouble-free production. Some transfers are air or hydraulic-cylinder operated; others are actuated by the movement of the ram or by independent motor drives. For heavy tonnage transfer presses where we must move and control many heavy stampings, the ultimate is the press driven combination cam and gear box feed. The feed drive takes its power directly from the main press drive.

The two most popular types are the single-axis or in-line feed and the tri-axis or lift and carry type feed. The in-line or single axis is applied whenever the stamping can be transported over relatively flat die surface. The stamping rests on the die surface and is grasped by the feed fingers and moved forward to the next die station. The feed bars move right to left only. Rack and pinion drives on the feed bars, through the feed finger boxes, cause the fingers to move in and out to engage the piece part and move it from station to station. This design permits the use of feed bars of any required length and will not generate vibration during the feeding action. The feed fingers are arranged for either electric or air sensing units to signal proper sequencing or loss of part conditions. Overload clutches are also provided to prevent damage to the feed or the tooling should a misfeed occur.

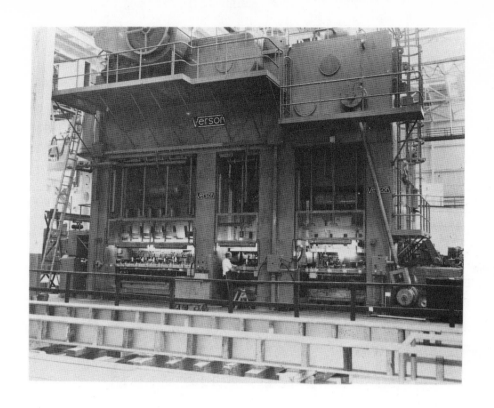

ILLUSTRATION NO. 5

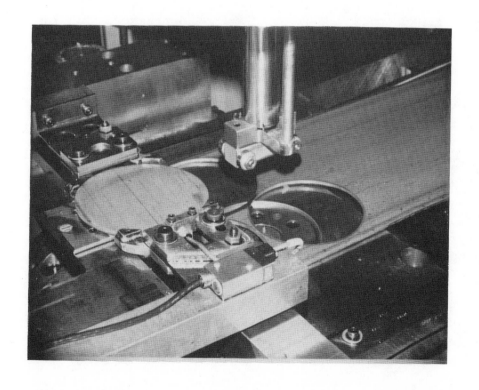

ILLUSTRATION NO. 6

ILLUSTRATION NO. 7

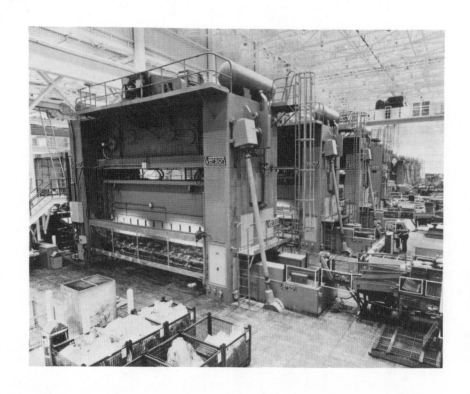

ILLUSTRATION NO. 8

The lift and carry or tri-axis transfer feed is a newer development. This very advanced and sophisticated mechanism guarantees complete piece part control at operating speeds not normally possible.

Illustration No. 8 is of a line of ten 1,800 and 1,000-ton tri-axis transfer presses which produce stainless steel automotive catalytic converters. Simply stated, the tri-axis transfer feed lifts the piece part above the lower die surface and transports it to the next die station and then lowers it on to the die. The feed bars are similar to the single axis bars except that a lower rack and pinion elevating mechanism raises and lowers the entire feed bars. The in-and-out finger motion is the same, plus gripping finger jaws which actually clamp on to the piece part during transport. By positive gripping, the speed of the feed can be increased due to the fact that the forward moving inertia of the stamping is totally stopped before the fingers release the stamping in the die position.

With a tri-axis feed, tool design can be greatly simplified. This is shown on Illustration No. 9. We are not concerned with guiding or sliding the stamping from station to station. The feed picks up, holds and transports and then positions the part in the next die and is not concerned with what is under the part as it moves.

ILLUSTRATION NO. 9

With these main elements of material feeding, transfer press, the transfer feed and electronic supervisory control, the total automatic system can now be established. With appropriate combinations of ancillary features and necessary controls, a total system will guarantee continuous self-maintaining automatic production. Some of these added features would be: (a) adjustable speed motor drives to permit maximum production at suitable press speeds for a specific stamping, (b) micro-inch drives to provide one-stroke-per-minute ram speeds during die set-up operations, (c) multiple automatic oil and grease lubrication systems for motors bearings, gears, gibways, feeds, etc., (d) pneumatic and positive knockouts in the ram over each die station for stripping and holding the piece part at the die level, (e) individual die cushions at each station as required for drawing or stripping purposes, (f) moving bolster units and automatic die clamping for rapid change-over of tooling.

The last major element of the total system transfer press is the electronic control package. Illustration No. 10 is a typical diagnostic control panel with elements of control in easy view of the press operator. Heavy tonnage presses of this type, operating at high speeds, necessitate continual surveillance, and the press operator or supervisor cannot observe or police all of the activities while the machine is operating. Consequently, electronic supervision is necessary. The development and sophistication of such control packages have now progressed to a very high level. The press when operating will automatically monitor all of the essential functions of the total system. Graphically, some of the points covered would be: (1) main drive motor load and temperature, (2) critical bearing temperatures, (3) lubrication system performance, (4) air line pressures and pressure devices, (5) coil and stock feed performance, (6) piece part location in each die station, (7) press tonnage monitoring, (8) overload systems monitoring of press, slide, feed clutches, etc. The future of press surveillance monitoring can go well beyond these essential points to the extent of total remote automatic production control and maintenance supervision.

As another example of a total system, Illustration No. 11 represents a 1,000-ton transfer press package with a press bed area of 6 feet front to back by 17 feet right to left. It has nine die stations on 22½-inch centers. The press stroke is 20 inches and operates from 4 to 40 strokes per minute. This press has the tri-axis lift-and-carry feed with the unique added ability to turn over the stamping as required to eliminate special cam dies. This one-press pressroom produces twenty-two different stampings. Frequently, there are two or three complete die changes per day. The automatic roller bolsters and quick die change features provided in the transfer system permit a complete die change in fifteen minutes. Presses with the design features providing such a total system can no longer be considered captive production machines and total utilization is possible.

ILLUSTRATION NO. 10

Probably one of the most elaborate transfer press systems ever put together was for a major manufacturer of automobile wheel covers. Obsolete equipment had to be replaced and production had to be increased. The original concept established from past experience by the customer called for a blanking press, a double-action drawing press and an automatic transfer press to perform the remaining stamping operations.

The wheel covers are stainless and the first draw operation requires the controlled slowed-down speed and high blank-holding capacity of a double-action press. The draw speed would dictate the production speed. To realize a minimum production of eighteen to twenty pieces per minute, two draw presses would have been required in the line. To circumvent this problem, the transfer press was designed to include all blanking, drawing and secondary operations in one machine.

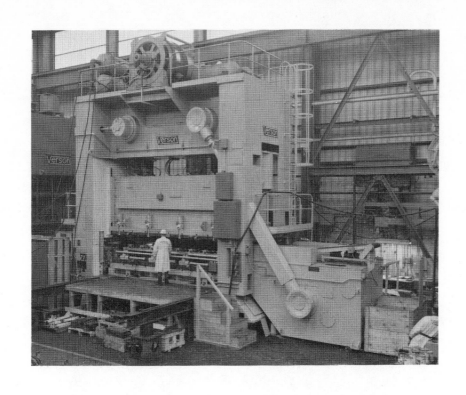

ILLUSTRATION NO. 11

ILLUSTRATION NO. 12

Illustration No. 12 is the 2,200-ton capacity result. A zig-zag grippper feed feeds a stainless coil to the sixty-ton blanking station which is outside of the right-hand column. The blanking ram is operated by the main press drive. With each press stroke, the coil is moved right to left and then front to back. The blank then moves to the first draw station. This is the 600-ton, single-point, 48-inch by 42-inch section of the press which simulates a double action. A one-hundred-and-seventy-five-ton blank holder cushion is provided in the slide and a one-hundred-ton die cushion in the bed, all of which operate together as a triple action draw press. In order to provide a safe draw speed, it is required to slow down and control the velocity of the ram as it performs the draw operation. To do this and still maintain an acceptable production speed, the press was equipped with a dynamic constant energy drive system. This adjustable speed drive permits a fast advance to a predetermined point, slow-down during the draw and return at a fast return speed. In this case, the advance speed would be 22 spm, reduced to 10 spm during the draw, then returned to 22 spm for a continuous productive output of about eighteen pieces per minute. The left-hand press ram is 1,600 tons and 4' by 19' right to left. There are nine die stations on 26-inch centers. The transfer is a tri-axis feed which was necessary to insure accurate positioning of the stamping in each die and which also made it possible for the user of this press to adapt and utilize existing tooling.

The past ten years have provided very significant developments of the transfer press. The industry has advanced from something of value to a definite necessity. Production capabilities make any other consideration for high-volume stamping production obsolete. This fact will insure the future and technological development of this type of metal forming machine tool. The transfer press concept is not only successful in metalforming operations, the transfer feed is also used extensively for high production of cold extrusion press work and die forging presses.

CHAPTER 3

INSTALLATION, MAINTENANCE & REPAIR

Press Isolators—Their Function And Effectiveness

By

SHELDON E. YOUNG, PRESIDENT

VIBRO/DYNAMICS CORPORATION

Originally presented at the Sixth International Congress on Sheet Metal Work, May 18, 1979

It is indeed a pleasure and an honor for me to address this congress.

The topic of my discussion today is PRESS ISOLATORS - THEIR FUNCTION AND EFFECTIVENESS.

For the past 14 years VIBRO/DYNAMICS has been engaged in a comprehensive study of press structures and the interrelationship between the press and foundation.

This study was made along three parallel paths:

(1) An analytical study of the forces acting upon the press structure and the behavior of the press structure

(2) A series of measurements of foundation forces, vibration and noise on a wide variety of presses

(3) Observation of thousands of presses mounted in every conceivable fashion.

When we started this program most everyone in the stamping industry believed that presses had to be bolted down and that isolators caused more problems than they solved. In many cases this was true.

However, we were convinced that if the isolators were specifically engineered for each press, to control the dynamic foundation forces and to eliminate the foundation twisting forces, it would be possible to achieve significant reductions in vibration and noise without causing harm to the press or dies.

Our studies have shown this to be true. What we did not expect when we started out was that when the isolators were properly designed and installed, the performance of the presses actually improved and very significantly. We also learned that vibration and noise could be reduced by a much greater amount than anyone had thought possible.

We also learned that hard mounting the press to the foundation is harmful to the press, to the foundation and to personnel; that by properly isolating the press these harmful effects are reduced.

With such isolators –

– Press performance and productivity are improved

- Downtime is decreased

- Repair costs are reduced

- Vibration and noise are decreased

- Die life is increased

- Power consumption is decreased

Our first introduction to the isolation of heavy presses occurred ten years ago. The 800 Ton press shown in Figure 1 had been bolted to a very heavy and wide foundation due to the wet sandy soil. In 18 months of operation the foundation had settled six inches. Adjacent building columns had also sunk requiring six inches of spacer plates under the building columns so that cranes would not have to run uphill. Cracking of foundation walls resulted in many thousands of dollars of repair work.

We were asked if isolators could be designed to isolate their new 800 Ton press in a way that would prevent similar problems.

After a thorough study of the stamping forces and the behavior of the press structure we concluded that it was not so much compressive or downward forces exerted by the press feet on the foundation that caused the foundation to settle.

Immediately upon snap through, the springback of the tie rods pulled up on the bed of the press and exerted a tensile force on the anchor bolts. These lifting forces were so great that they raised the foundation causing water in the soil to be sucked into the area under the foundation. The subsequent pumping action as the foundation vibrated caused sand and silt to wash out from under the foundation.

We concluded that one of the primary design objectives of these press isolators was to prevent the press from exerting an upward force on the foundation. A second design objective was to reduce the downward impact force on the foundation. A third objective was to provide a simple means for leveling the press with a higher degree of accuracy than could otherwise be attained. A fourth objective was to design the isolators to have a long life without deterioration.

Both presses were installed on these isolators. In these ten years neither foundation has settled, no further damage has occurred and the presses have remained level.

Additional benefits were observed. At the operators station noise measurements showed a 6.5 dB reduction (Figure 2). In the frequency ranges that are most damaging to the human ear, reductions up to 12 dB were measured. Vibration in the floor was also reduced.

In 1971 the press in Figure 3 was bolted down to a foundation. However, it experienced a great deal of downtime and could not be operated at full design speed of 300 SPM. It was then reinstalled on press isolators.

Last week I talked with the man in charge of die design and the man in charge of press installations for this company. They reported that on all 8 of these presses the isolators--

- Increased die life 600 to 800%

- Increased operating speed from 270 to 300 SPM

- Decreased vibration and noise very effectively

- Decreased press downtime

They also reported that both the press and the press isolators are performing as well today as they did in 1971 when the isolators were first installed.

It is significant to note that no one knew that the cause of the problem was the manner in which the presses had been installed until the problem was solved by the use of isolators. Until that time there had been a series of confrontations between the press user, the press builder and the builder of the transfer feed and die system.

By working closely with all three parties we were able to design isolators which provided optimum performance for these specific presses.

These problems are universal, regardless of the type of construction of the press or who the manufacturer is--and we have seen the same beneficial effects when the press has been properly installed and fine tuned on isolators that have been custom engineered.

The importance of optimizing the design of isolators is illustrated by the 400 ton cold forging press in Figure 4. The plant was laid out for maximum efficiency of material handling with the heat treating furnace located about 20 feet from the press. Realizing that vibration might be transmitted to the furnace unless something was done to prevent it, the press was mounted on a heavy foundation that was "isolated" from the surrounding floor by a resilient separator; and a pad of resilient material was placed under each of the press feet to isolate the press from the foundation.

The main item of concern was the sensitivity of the mercury switches that control the heat treating furnace. These switches had been tested to determine that vibration in excess of .0015 inch would affect their reliability.

When the press was operated the heat treating furnace could not be operated. Vibration measurements showed that the mercury switches vibrated with an amplitude of up to .014 inch

(9.3 times as much as the switches could tolerate).

The resilient pads were removed and a set of leveling isolators was installed. Vibration of the switches was reduced to .0015 inches.

Inasmuch as this was still a marginal condition, a third set of isolators was installed. These isolators were designed to provide optimum reduction of the foundation impact forces and to eliminate the foundation twisting forces. As a result vibration of the mercury switches was reduced to .0007 inch.

With the resilient pads supporting the press the vibration transmitted to the mercury switches was 20 times as great as it was when the press was installed on the optimum isolators.

Other effects were observed. When the press was first mounted on resilient pads, excessive vibration was transmitted through the soil to other parts of the building and to other buildings. Quality control work had to be scheduled between shifts. Vibration and noise in an office building 150 feet from the press were intolerable.

With the press installed and fine tuned on the optimum isolators, quality control functions were unaffected by the operation of the press. In the office building it was not possible to detect when the press was operating and when it was not.

In our study we felt that it was important to determine more precisely how effective isolators can be in controlling the foundation forces and the reduction in vibration and noise that results from the control of these forces.

Our analytical study showed that for most presses it is possible to design isolators that reduce foundation impact forces and foundation vibration by 95 to 98%.

These computed values of effectiveness were substantiated by actual measurements in the field. This 250 Ton press (Figure 5) was bolted to a foundation but it was decided to reinstall it on press isolators because of the harmful vibration which was being transmitted to other equipment.

Prior to reinstallation a series of vibration measurements were made on the press structure, on the foundation, on the floor, and on the building column. Similar measurements were made after reinstallation of the press on isolators that were designed for optimum control of dynamic and static foundation forces.

Overall vibration of the foundation was reduced by 31 dB as indicated by the vibration meter. This is equivalent to a reduction of 97.2%. Figure 6 shows a typical set of measurements. Both halves of the figure are identical. The left hand side shows the vibration as measured in dB. The right hand side shows the same vibration defined in G's. Both sides show that the isolators reduced vibration by 97.2% which is another way of saying 31 dB attenuation. The left side shows something else that is very significant; that when the press was installed on the isolators the noise radiating from the foundation surface was attenuated by 31 dB.

Putting it the other way around, when the press was bolted down, the vibration in the foundation was 36 times greater than it needed to be and the noise radiating from the foundation was 31 dB higher than it needed to be; that the isolators prevented this excess vibration and noise radiation.

Figure 7 shows that similar measurements showed that the

press leg vibrated 7 times greater when the press was bolted down that when it was properly isolated and that the equivalent noise radiation from the leg was 17 dB higher with the press bolted than with it isolated.

Vibration of the floor was 10 times greater and radiated noise from the floor was 20 dB greater with the press bolted down (Figure 8).

Vibration of the building column was 6 times greater and radiated noise from the building column was 16 dB higher (Figure 9).

One very personal side benefit of this improvement showed up when the press operator expressed his appreciation to the plant manager inasmuch as he no longer went home fatigued. Not only was his performance improved at the shop. We understand from his wife that his performance at home also improved.

As a noise control measure properly designed press isolators are unique. While most other noise control measures tend to interfere with productivity, and increase costs, isolators increase productivity and decrease costs.

If we look at a press resting on its foundation we see that it tends to be supported under three feet due to irregularities in the surface of the foundation (Figure 10). You will notice that the shape of this floor is such that the left front foot is unsupported. Most of the press weight is supported on the left-rear/right-front diagonally opposite feet.

As a result the press bed is subjected to twisting forces. These twisting forces cause the bed to warp into the shape of a saddle and the entire press structure to twist about a vertical axis

(Figure 11). Columns and tie rods are caused to bend and twist.

Press isolators which are properly designed to reduce the foundation impact forces and to eliminate the twisting static forces incorporate a very precise adjustment means for compensating for the variations in elevation of the foundation under the four feet.

The press is leveled by turning the precision adjusting screws in the four isolators until the press is both level and flat and so that the press supports are perfectly fine tuned. (Figure 12)

The four-channel oscillograph illustrated in Figure 13 shows the weight supported by the four isolators under a 60 Ton press.

When this press was first set down most of the weight was supported by the left-rear and right-front feet because the elevation of the floor was higher under these two feet. The precision adjusting screws in the four isolators were adjusted to do two things:

1. To precisely level the press bed

2. To remove all twisting forces caused by floor irregularities

This is called fine tuning. The adjusting screws were turned to transfer the excessive load supported by the left rear and right front isolators to the left front and right rear isolators until 50% of the weight is supported by each of the two diagonal pairs of isolators.

The press shown in Figure 14 was leveled and fine tuned in 10 minutes resulting in a reduction in power consumption. Reductions in power consumption of high speed presses of 5 to 25% are not uncommon.

Perhaps the best way to summarize the function of isolators is this:

They permit the press to perform to its full

capabilities by preventing the foundation from exerting

harmful influences -

- on the press

- on the foundation

- on the surroundings

and in so doing, they also improve the environment for

personnel and for other equipment.

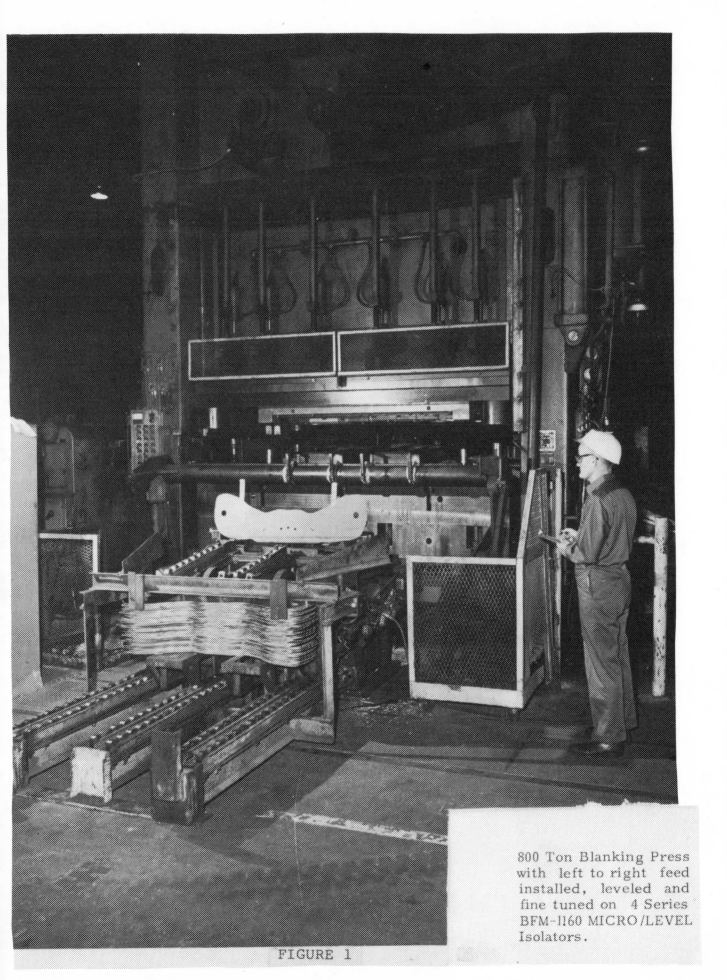

800 Ton Blanking Press with left to right feed installed, leveled and fine tuned on 4 Series BFM-1160 MICRO/LEVEL Isolators.

FIGURE 1

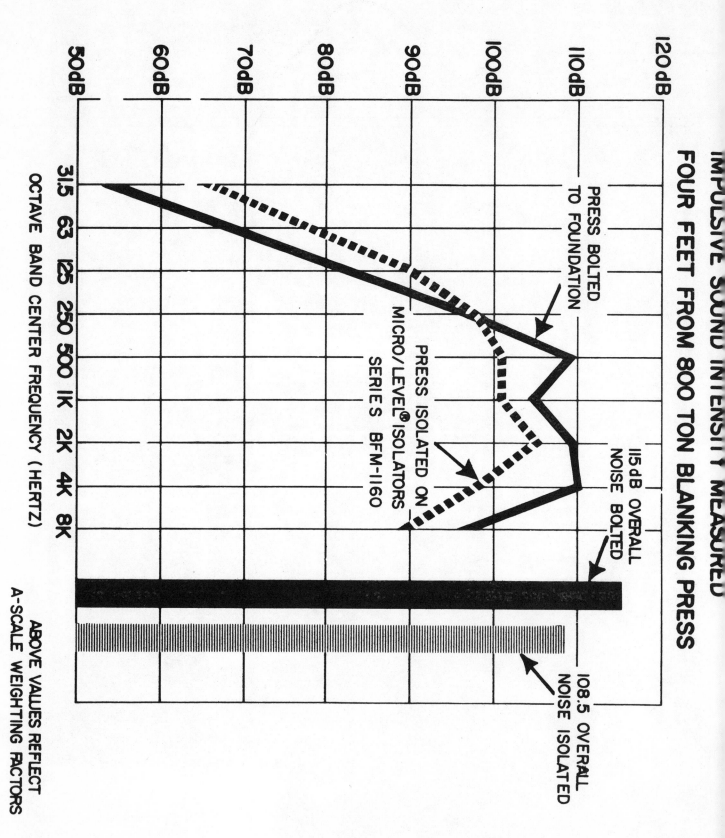

IMPULSIVE SOUND INTENSITY MEASURED
FOUR FEET FROM 800 TON BLANKING PRESS

PRESS BOLTED TO FOUNDATION

PRESS ISOLATED ON MICRO/LEVEL® ISOLATORS SERIES BFM-1160

115dB OVERALL NOISE BOLTED

108.5 OVERALL NOISE ISOLATED

OCTAVE BAND CENTER FREQUENCY (HERTZ)

ABOVE VALUES REFLECT A-SCALE WEIGHTING FACTORS

FIGURE 2

*INSTALLATION OF A NIAGARA PD2-35W PRESS WITH WILLETT TOOLING ON 10L600 MICRO/LEVEL ISOLATORS.

FIGURE 3

FIGURE 4

FIGURE 5

EFFECTIVENESS OF SERIES BFM-1230 MICRO/LEVEL ISOLATORS IN REDUCING VIBRATION CAUSED BY 250 TON BLANKING PRESS

VIBRATION IN RIGHT FRONT FOUNDATION

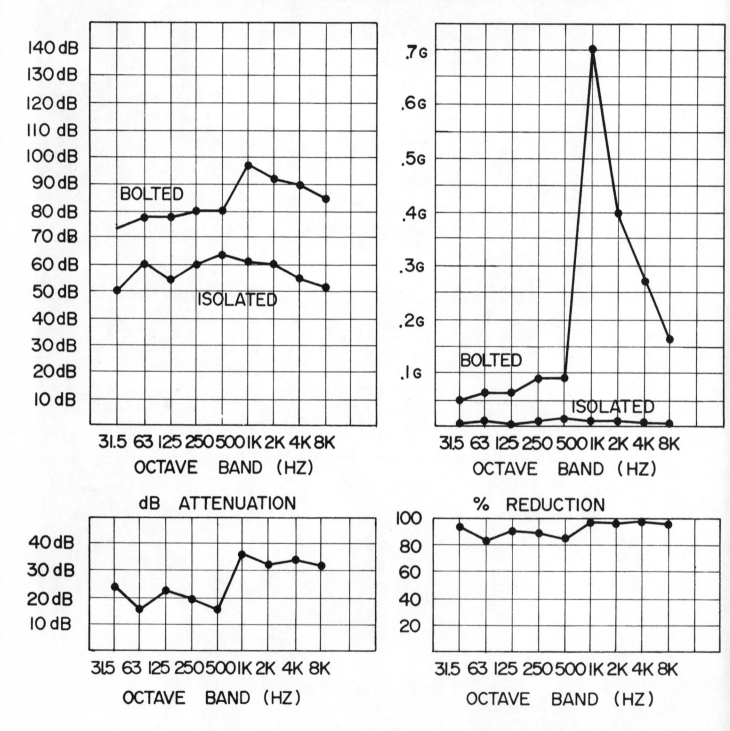

FIGURE 6

EFFECTIVENESS OF SERIES BFM-1230 MICRO/LEVEL ISOLATORS IN REDUCING VIBRATION CAUSED BY 250 TON BLANKING PRESS

VIBRATION IN RIGHT FRONT PRESS FOOT

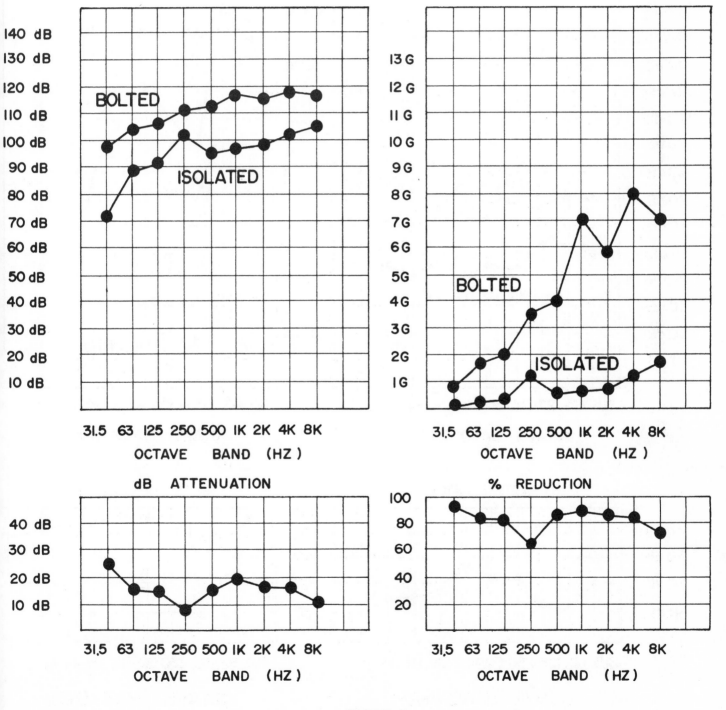

FIGURE 7

EFFECTIVENESS OF SERIES BFM-1230 MICRO/LEVEL ISOLATORS IN REDUCING VIBRATION CAUSED BY 250 TON BLANKING PRESS

VIBRATION IN CONCRETE FLOOR

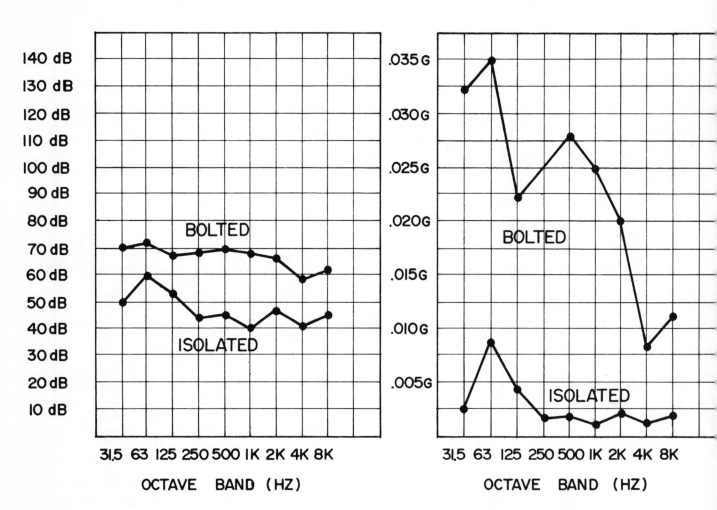

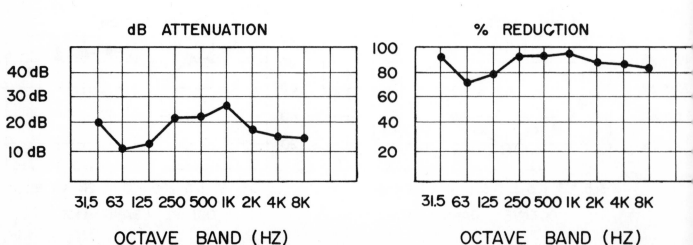

FIGURE 8

EFFECTIVENESS OF SERIES BFM-1230 MICRO/LEVEL ISOLATORS IN REDUCING VIBRATION CAUSED BY 250 TON BLANKING PRESS

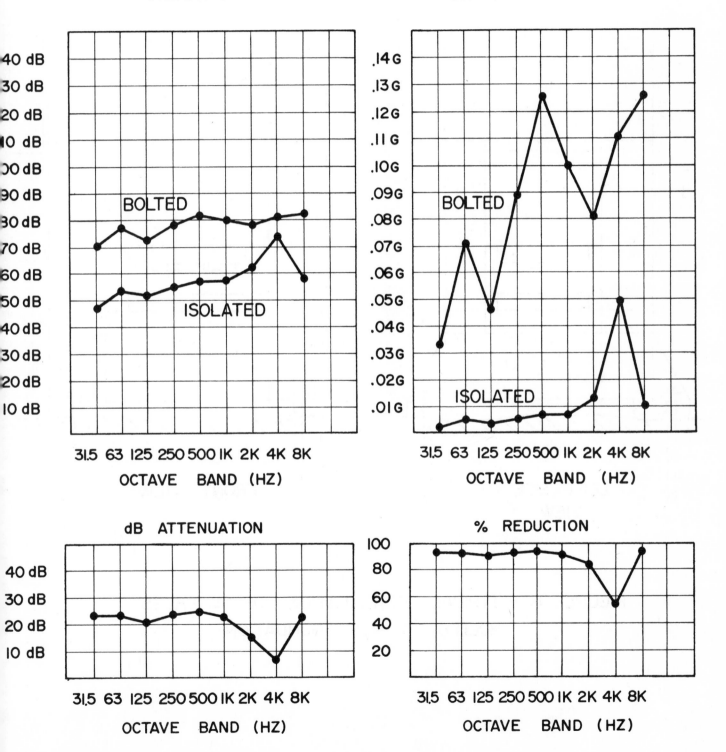

FIGURE 9

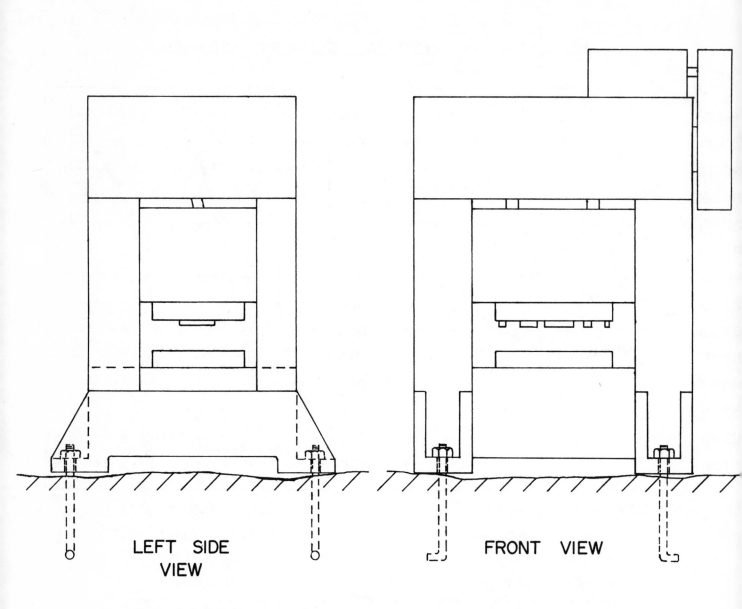

LEFT SIDE
VIEW

FRONT VIEW

FIGURE 10 - TYPICAL INSTALLATION OF A PRESS
BOLTED TO ITS FOUNDATION

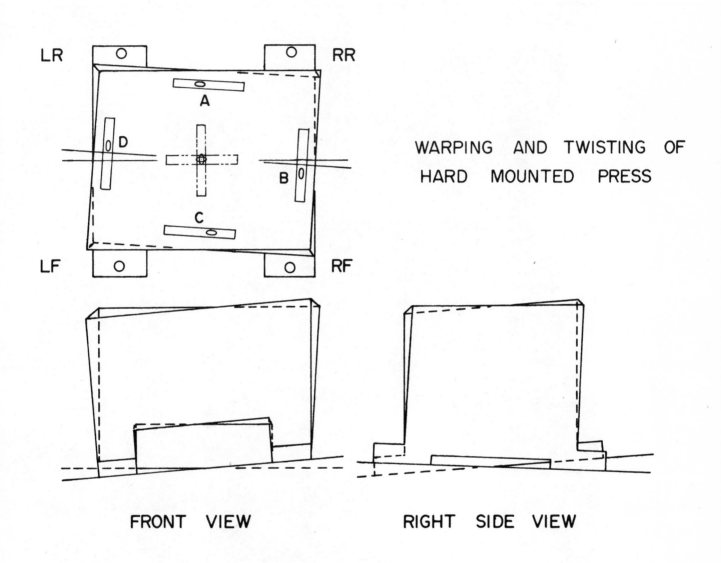

LR RR

A

D

B

C

LF RF

WARPING AND TWISTING OF
HARD MOUNTED PRESS

FRONT VIEW   RIGHT SIDE VIEW

FIGURE 11

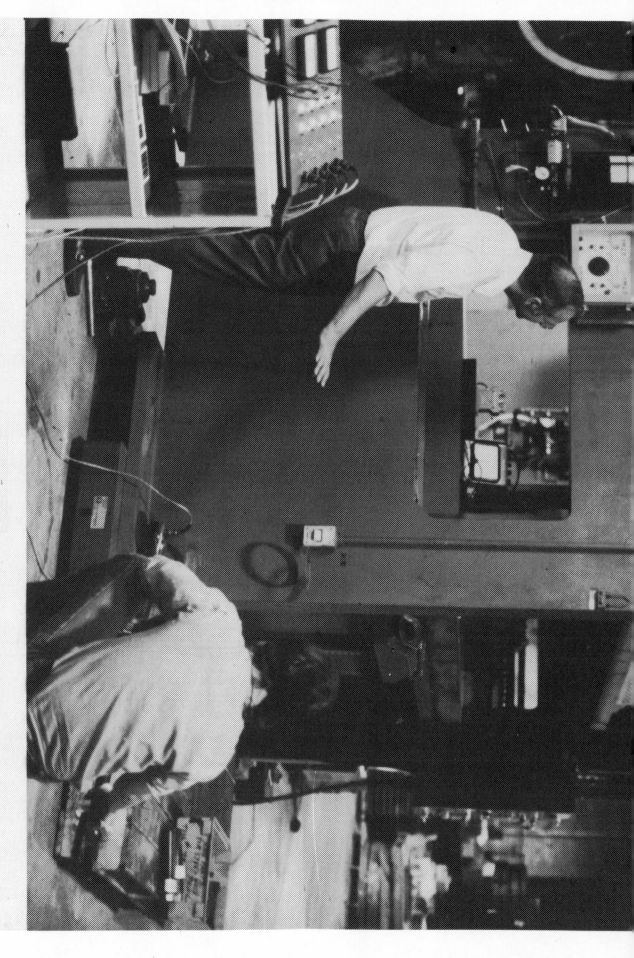

FIGURE 12

102

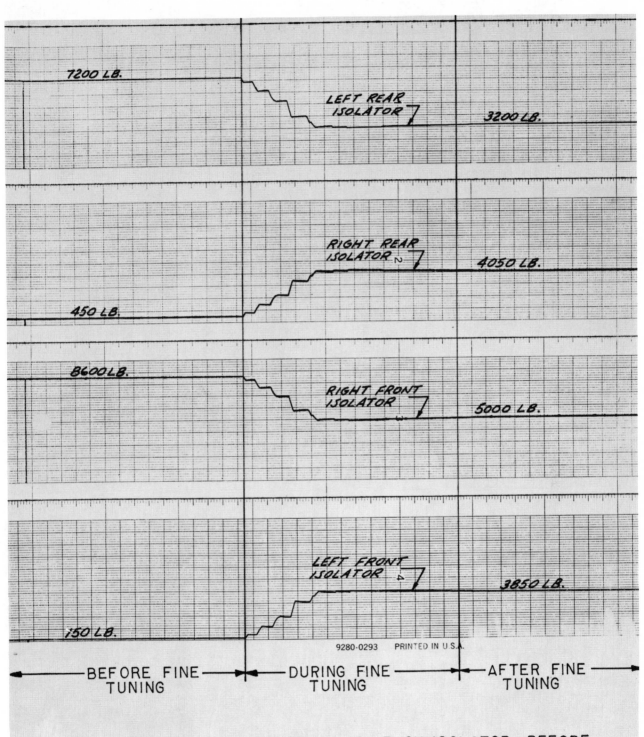

BEFORE FINE
TUNING

DURING FINE
TUNING

AFTER FINE
TUNING

FIGURE 13 - WEIGHT SUPPORTED BY EACH ISOLATOR - BEFORE
AND AFTER FINE TUNING 60 TON PRESS WITH
LOAD SENSING ISOLATORS.

FIGURE 13

FIGURE 14

104

Power Press Noise Reduction

You don't have to be an acoustics expert to tackle power press noise problems. Focus your engineering knowhow on four major noise sources for significant improvements.

By JAMES H. CAPPS
*Corporate Consultant—Safety
General Electric Co.*
Fairfield, Conn.

If your plant uses more than a few power presses, the chances are you have a noise problem. A quick test is to try to carry on a normal conversation at an operator's station. If you cannot, the noise level is probably above 90 dBA—in which case, you have a noise problem!

There are significant costs associated with noise at that level, such as the cost of hearing protectors and mandated audiometric testing. And, let's face it, although continuing exposure will damage hearing, employees do not like to wear protectors, so those devices do not eliminate the possibility of hearing damage and consequent compensation claims. The solution to the problem is to reduce the noise.

Press room noise has many causes. Some noise can be traced to easily identifiable singular sources; other noise has no single source, rather it is general, resulting from a combination of causes. Much of the noise can be reduced for a relatively small cost and with relatively little engineering effort.

The general strategy is to identify the singular noise sources and

* The byline reflects the author's position at the time material for this article was developed. He is currently Safety Engineer, Div. of Safety Research, National Institute for Occupational Safety and Health, Morgantown, W. Va.

attack them, worst first, in a series of development projects (except that any noise cause which is not relatively easily understood should receive a lower priority).

Note that this strategy focuses on the source of noise radiation and not on such variables as a force, amount of vibration, internal damping, resonance, etc., that a noise reduction consultant would deal with in solving a knotty problem.

Identifying the source of noise in power presses is generally not a difficult problem. In this suggested engineering approach, identification of sources does not require sophisticated analyses of frequency and precise noise levels for at least a major portion of the effort. Some measurements with a simple sound level meter may be useful to confirm progress, but beyond that, efforts to measure noise levels or characteristics are not required.

What is required, however, is a knowledgeable person—or persons—with a good pair of ears.

Typically, the singular sources of noise occur at different times in the press operating cycle. The experienced engineer with a "good pair of ears" (and, maybe, the sound level meter) can note the timing of the noise or the ability to control the timing to separate, and identify, the noise sources.

Four sources can be distinguished:
■ Material handling noise (exterior to the press).
■ Press equipment noise (motor, clutch, control air exhaust, etc.).
■ Die space noise (stripper plate, pin impact, etc.).
■ Press structure noise.

The beginning point, then, for an engineered attack on a noisy press room is an understanding of the noise from these four sources and an appreciation of how the noise can be reduced or controlled. A step-by-step attack on the four sources will yield reductions up 10 to 20 dB at probably a majority of the workstations.

Handling noise

Material handling noise is associated with getting work material into and out of the press. Obviously, noise of this sort varies widely, depending on the process, the parts produced, and the method of handling. The noise is radiated from the part or material itself or, in some cases, from the material handling equipment. It can arise anywhere in a press room.

Separating material handling noise from other noise is probably not difficult. The task is identifying the more significant sources of this kind of press room noise. The judgment of a couple of observers will provide enough information for initial action. Occasionally, it will be necessary to operate a piece of material handling equipment at a time when no other equipment is operating in order to judge its effect.

Reductions of material handling noise depend on the simple tactic of "change the energy change." For example, change the velocity of the impact (don't drop the material so far) or change the time of deceleration (cushion the impact) or prevent the impact of part-on-part.

Here are some examples: Automobile control arm blanks, about a quarter of an inch thick, and weighing ten pounds, were being dropped six feet into a metal tote bin at a fairly high repetitive rate from a roll-fed press. As an improvement, the parts were slid into a cushioned rack with considerably less noise and a significant cost saving at the next operation because the parts did not need to be retrieved from the bin.

A blank for a truck frame side rail was ejected from a press onto a bed of steel roller wheels with the wheels so spaced that the frame blank bounced from one group of wheels to the next. The fix is to change the type of conveyor or perhaps to change to plastic or rubber wheels to avoid the sharp

impact (presuming a material could be found which would withstand the rough duty). This noise was obvious over a fairly large area of a building, so that its reduction would improve the quality of a large work area.

Equipment noise

Press equipment noise has its source at the press. On newer presses, this if often not a significant noise. But some drive motors, clutches, control air exhausts, etc., are causes and may be significant, particularly in combination with other equipment in the press room. For large forming presses, gear noise may be a large contributor.

Press equipment noise is easily recognized but turning equipment on and off may be needed to separate noises from different presses. The exhaust from pneumatic controls may be singularly loud or noticeable but may not contribute significantly to overall press room noise. However, occasionally, the combination of air exhaust noise of a number of presses may be significant in a press room. For gear noise, cycling a press without its die set may be instructive, although often gear noise depends upon the loading provided by the press working.

Reducing press equipment noise generally involves an attack at the cause—which means either change the configuration or improve the maintenance status of the equipment.

For example, air exhausts can be quieted by adding commercially available mufflers. Their effectiveness can suffer because of loading up with oil mist or other materials but this problem can be attacked by adding filters to the supply lines—which is wise in any case since contaminated air also causes other problems with the pneumatic controls.

Occasionally a totally enclosed fan cooled drive motor is found to

Stripper plate impact— a significant noise source

If your die set includes a stripper plate, that component is a likely source of die space noise. In the typical design, the plate is free to vibrate. And note from the graph that the stripper plate contacts the work at a relatively high velocity. In contrast, the impact of the punch with the workpiece generally is not a source of noise because it occurs near the bottom of the stroke (lower velocity) and the workpiece is somewhat resilient. Die space noise is often confused with press structure noise in high speed presses because of the small time separation between the two noises.

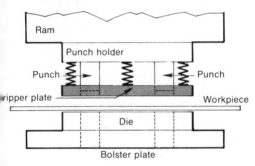

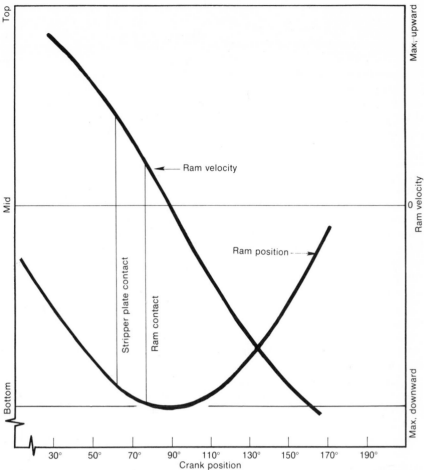

be a major source of noise. Changing a motor is a major expense and should be considered only in unusual circumstances. Conducting the motor ventilation exhaust away from the work area may be helpful but should be tested by first building a rudimentary trial structure of some sort.

On forming presses, significant gear noise often is the result of wear. The noise may portend an early failure and thus the gears should be replaced even though they represent a major expenditure. Another possibility is that the gear housing may be a major noise radiator. Damping or strengthening this structure may be quite helpful. Here again, some sort of simple tryout should be devised before an expenditure is made for a change.

Die space noise

Die space noise is often assumed to be the major noise source in a press. While this may be true for a few applications, more often than not the predominant noise has other sources.

Noise emanating from the die space can generally be classified into that caused by the stripper plate (if used) by die secondary motion such as restrikes or by knock-out impacts, or by material handling, such as dropping parts on a shuttle.

Stripper plate impacts may be a significant source of noise because, at the time of contact, their velocity is still relatively high and because stripper plates are often mounted so that they can vibrate like a free beam. Secondary forming and other die auxiliary motions are often designed to have high impacts which cause the work material and parts of the die to vibrate and radiate significant amounts of noise.

Here, note that the impact of the punch with the die or workpiece generally does not cause a significant amount of noise. The impact occurs near the bottom of the stroke so that the velocity of the punch is relatively low. Also, the contact with the workpiece is relatively soft (the workpiece is somewhat resilient). The noise that is associated with this imagined impact derives from another source and is categorized as press structure noise.

Identifying and separating die space noise from press structure noise usually requires operating the press under various conditions. One condition, which is usually easy to achieve, is cycling the press with the die set in place but without advancing the work material. With the die set in place, the impact of the stripper plate, the restrike cams, or the knock-out pins will occur in essentially the normal fashion. Since work is not being done, there is no press force developed—therefore, no press structure noise. And, of course, there is no material handling noise emanating from the die space.

Die space noise reductions are often somewhat difficult to obtain. Their reduction can depend on the philosophy used for die designs. Thus, even though future dies may be significantly quieter, existing not-so-quiet designs may be in use for a number of years. The latter problem suggests the use of something like enclosing the die space.

On high speed blanking presses, the most common die space noise is stripper plate noise. As noted, stripper plates are mounted in such a way that they are relatively free to vibrate. Internal damping or cushioning on the lower edge of the stripper plate can be provided by careful redesign. However, stripper plate applications are rough duty so that cushioning or internal damping often fails.

Similarly, restrike and knock-out pin impacts can be reduced by cushioning—but, again, they are rough duty. Impacts of workpieces falling onto a shuttle after being dropped from the punch section of the die set can be controlled through design changes.

A die space enclosure can be considered when not precluded by material handling considerations. Also, enclosures have to be designed with full recognition of the operator's need to observe the punching or forming action, and clearing of scrap. Whatever, enclosures can be attractive, problem solutions achieving a reduction of 10 dB or more where die space noise is predominant.

Before an enclosure is purchased, the concept should be tested by building an elemental enclosure of the die space.

Structure noise

Press structure noise has its origins in the forces developed during the punching action of a press. Of most significance is the rapid reduction in force which occurs in a punching operation at the point of breakthrough. At this point, the maximum punching force is rapidly unloaded. Since these forces are distributed throughout the press structure, the Pitman, the drive shaft, top hat, etc., any part of the press which is "tuned to" and can react to the rapid force reduction will do so by vibrating in the audible frequency range and thus causing noise. This noise depends upon the load on the press, the inertial effects of the whole punch and die support system, and the capability of various surfaces to react—which involves their resonance, internal damping, etc.

Cycling a press with and without the work material advancing will separate die space from press structure noise. More specifically, if a major reduction in noise occurs when the material is not advanced, the noise is invariably press structure noise.

Reduction of press structure noise is the most difficult to achieve. It tends to be the predominant noise for high-speed blanking presses which punch relatively hard material. It is often radiated from the upper portions of the press frame (the top hat).

Most proposals for structure noise reduction involve reducing the force required during the punching operation. For example, grinding shear on the die will reduce it. If the shear halves the force required, the noise is reduced about 3 dB. Punch clearance and other changes of this type also have the same effect. Occasionally these changes are practical, but they are not universal.

Reducing the rapid unloading of the press frame on breakthrough is another possibility. Shock absorbers have been designed and can be shown to be effective. However, they are complicated to build and costly to maintain. Their point of action must be tuned to the point of breakthrough. The forces and the stroke rate are high (at the press rating), which means that there is a large amount of energy to be dissipated. Commercial shock absorbers for small presses are offered for sale by Hylatechnic, a German supplier. No United States suppliers are known. The use of shock absorbers may reduce the noise 10 to 15 dB, but their widespread application awaits development.

Die cushions may offer some reduction in noise, but they do not do more than eliminate load reversal which normally occurs.

In recent years, press structure designs have been changed by adding internal damping or stiffening to reduce the response to the rapid force reduction. If, for an older press, the top hat is found to be the

---

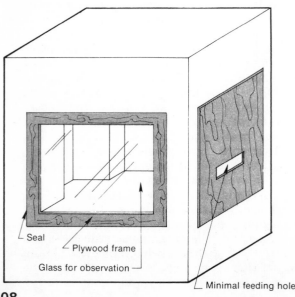

A trial enclosure for a die space

It's a good practice to test the possible effectiveness of a die space enclosure with some simple version before obtaining a final form. The trial enclosure can be constructed using ¾-inch plywood and weather stripping as a seal. Thermopane glass should be used in creating an observation window. In addition to testing noise reduction effectiveness, a trial enclosure enables operating and maintenance problems to be evaluated.

Seal
Plywood frame
Glass for observation
Minimal feeding hole

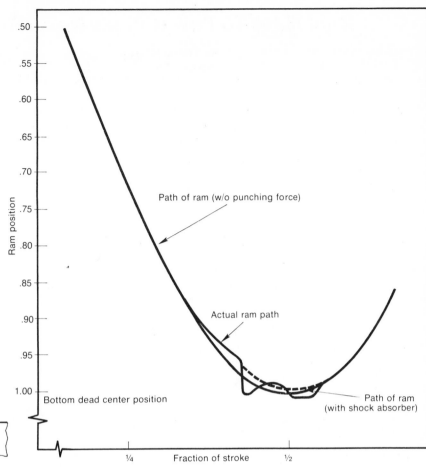

Shock absorbers can reduce press structure noise

Although not widely used, shock absorbers can be shown to be effective in noise reduction by reducing the rapid unloading of the punching force in a press. Not only must the devices dissipate a large amount of energy, they must also be tuned to the point of breakthrough. They are complicated to build and costly to maintain.

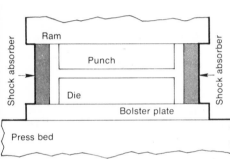

major source of the noise, reductions of 3 to 6 dB might be achieved by rebuilding. A partial enclosure—that is, an enclosure limited to the surfaces radiating the noise—may also provide a 3 to 6 dB reduction, and may be practical for some installations. Here again, a trial of test enclosure is needed to confirm the reduction potential before a significant expenditure is made.

Total enclosures

Despite the general utility of a strategy of attacking the four sources of noise, it is likely that there will be a few workstations involving a few presses for which the approach is not effective. In these cases either the press structure noise is predominant or the structure noise in combination with other types of noise produces a problem so complex that analysis cannot be made effectively.

For these presses, total enclosures should be considered. Enclosures are feasible for high speed blanking presses where the input and output material handling is not severely limiting and if there is space to accommodate the enclosure.

Enclosures provide reductions of 15 to 25 dB or more if they are well-designed—and well-accepted by the operators and maintenance technicians. Commercial designs are available, but caution is advised in their application. The feasibility study should be made by a team consisting of an acoustical consultant and a responsible manufacturing engineer well versed in the problems of maintenance and usage of such a structure.

For press rooms without singular noise sources but a generally uniform, high noise level, wall treatments and ceiling treatments may be useful. Reductions of approximately 3 dB can be achieved by a careful design of the wall and ceiling treatments. Again, an acoustical consultant is needed.

There are other kinds of devices which are often suggested as means of press noise reduction. An example is the use of press vibra-

tion mounts. Although occasionally effective, such devices often do not have a significant effect on the noise at the operator's station.

The approach and techniques which have been discussed will substantially reduce the risk of damage to employee's hearing (often equated to OSHA compliance). And since the techniques are based upon current state-of-the-art of power press noise reduction, they are a good basis for action which will avoid a significant OSHA citation.

As to cost effectiveness, there is a cost reduction for each employee removed from a hearing conservation program (some corporations estimate a savings of $100 per person per year). And while studying a noise reduction problem, general press room operations should be examined for traditional cost reduction opportunities that can be accomplished at the same time. If these two sources do not yield a sufficient saving, employee satisfaction, avoidance of major OSHA citations, and similar considerations can be factored into the cost study.

Presented at the 1980 International Engineering Conference, April, 1980

Guidelines To Power Press Noise Reduction

By Lewis H. Bell
Vice President
Harold R. Mull, Bell and Associates (HMB)

Anyone who has dealt with power press noise control knows there are no quick and easy solutions. Worse yet, each press and/or tool set usually must be dealt with on an individual basis. This is due primarily to the number of complex noise sources, insidious noise paths, production constraints, etc. In short, there are no "gimmicks" or sweeping measures available. However, significant progress has been made and it is the object of this paper to:

1. Identify, characterize and rank order the major sources of press room noise; and

2. Present state of the art developments and design guidelines for noise reduction.

In addition, the typical acoustical performance and/or limitations of each noise reduction measure is also given. In this way, a systematic noise control program with realistic goals can be established and implemented with cost effectiveness assured.

INTRODUCTION. Power press rooms have over the years been a formidable challenge to the acoustical engineer. There are dozens of noise sources, some of which can be traced to easily identifiable singular causes while others result from a complex combination or sequence of events. In addition, the noise paths are, in some cases, straight forward and direct which others are insidiously confounding. However, recent press noise research and extensive development programs have identified the major causes of noise and some effective noise reduction measures have been developed and are now available.

It cannot be overemphasized however, that effective treatment and significant noise reduction does not come easily. To be sure, only by systematic and careful attention to detail from the design of the tool to collecting finished parts is there a chance for any measure of success.

It shall not be the objective here to provide specific design details for press room noise control, but rather to provide the tool design and manufacturing engineers, with some basic guidelines for noise reduction which will apply to many press room operations.

MAJOR SOURCES OF NOISE. As a starting point, let us consider the four areas or functions which are the major sources of noise. In approximate order of noise intensity or "loudness" we have the:

1. Die space area; (tool impact, breakthrough fracture, stripper plate impact, and high velocity air ejection);

2. Press equipment; (motors, clutch, pneumatic control exhaust air, gears, etc.);

3. Press structure; (noise radiated from the vibrating structures, i.e., frame, flywheel, cover guards, etc.);

4. Material handling; (parts and scrap).

With these sources and areas identified, let us now consider each area individually, discuss the character of the noise and for each, the available noise reduction measures.

DIE AREA. Noise emanating from the die area generally originates from the following causes in somewhat sequential order:

1. Stripper plate (if used) impact with the work piece;

2. Tool impact with the work piece;

3. Breakthrough (fracture);

4. Restrike or knockout; and/or

5. High velocity air to eject parts.

Taking each source individually:

Stripper Plate: In high speed blanking operations, noise from the stripper plate is often the dominant source of noise. The only reported noise reduction measure is to "cushion" the plate where it strikes the work piece with some rubber or polyurethane pads or "snubbers" as illustrated in Figure 1a below. Since the impact of the plate is a punishing "heavy duty

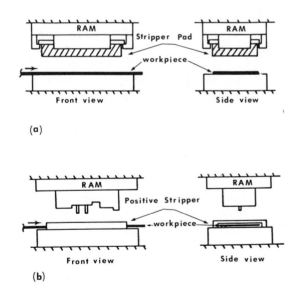

(a)

(b)

FIGURE 1   Illustration of a) acoustically treated stripper bars and b) positive stripper configuration.

cycle," the life cycle characteristics and integrity of materials attached to or imbedded in the stripper plate is typically brief, and, as such, a high maintenance item. In short, these materials wear or deform quickly and require frequent replacement. Treatment of this type applied to knockout pins has also been shown to be troublesome.

Application of viscoelastic damping material to the stripper plate or die set has also been widely reported. However, there is also a preponderance of data that suggests there is little, if any, noise reduction from this superficial treatment.

As an alternative to the stripper plate, a "positive stripper" approach eliminates completely the stripper plate impact noise. The positive stripper when applicable is shown conceptually in Figure 1b. Here, the parts are stripped with little or no impact noise.

Tool Impact with Work Piece: To understand the tool impact and break-through noise, consider the following sequence of events. As the tool impacts the work piece, a tremendous amount of energy is expended over an extremely short period of time. Even though only a small part of the expended energy is converted to sound, the result is a high intensity, impulsive-like report. Peak sound pressure levels from the punch or tool impact are often in the range of 130 to 150 dB. The noise as the tool breaks-through the work piece is also impulsive in character. The magnitude of the noise level depends to the first order on material hardness, thickness, etc. Significant noise level reduction for these sources is difficult to achieve, but two factors are fundamental:

1. If the duration of the impact can be extended, and

2. If the speed of punch can be lowered, the noise levels are generally reduced.

To achieve these goals the following measures have been shown to reduce tool impact noise where applicable:

1. Utilize shear, tapered, or step-cut dies or punches as illustrated in Figure 2.

2. Utilize a press with sufficient or extra capacity and minimize ram stroke speed.

With respect to punch modification, the shearing action tends to extend the impact duration and/or ease the metal cutting action. Intuitively, a slower ram stroke with equivalent force reduces the punch speed and also extends the impact duration.

Breakthrough or fracture noise appears rather unique in that no noise reduction measures or even research on this source have been published.

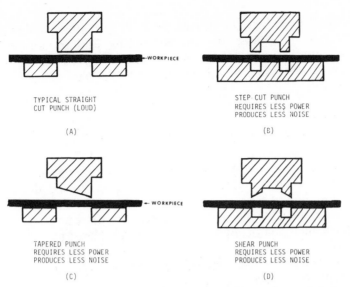

TYPICAL STRAIGHT
CUT PUNCH (LOUD)

(A)

STEP CUT PUNCH
REQUIRES LESS POWER
PRODUCES LESS NOISE

(B)

TAPERED PUNCH
REQUIRES LESS POWER
PRODUCES LESS NOISE

(C)

SHEAR PUNCH
REQUIRES LESS POWER
PRODUCES LESS NOISE

(D)

Figure 2   Illustrations of die set modification which reduce noise.

It is important to note, that studies relating to treatment of stripper
plates and the use of shearing tools yielded noise reduction in the range
of only 2 to 5 dBA.   However, the use of larger capacity presses, than
necessary, can reduce radiated noise 10 dBA or more.   Unfortunately, some
production penalties are often incurred with slowing stroke speed and
costs for higher capacity equipment are not ignorable.

High Velocity Air to Eject Parts: In many automatic press operations, the
noise associated with high velocity air ejection nozzles is dominant. Noise
levels from a 1/4" diameter copper tube "crimped" to blow off a small part
are generally in the range of 98 to 100 dBA at 3 feet.   See Figure 3.

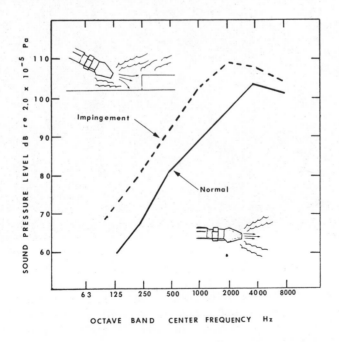

Figure 3   Typical noise levels associated with air ejection nozzles.

The sound power of a small high velocity air jet used to blow off parts varies as the 6th to 8th power of the velocity and as such is one of the most dynamic functions in physics. Correspondingly, noise levels may be sharply reduced by lowering the ejection air stream velocity. This is best accomplished by:

1. Moving nozzles closer to part or scrap being ejected and lowering velocity. Halving the distance between part and nozzles will allow air velocity to be lowered 30% (preserving thrust) with a noise reduction of 8 to 10 dBA.

2. Adding additional nozzles with lower air stream velocity, again preserving thrust.

3. Using recently developed quieter eductor/diffuser nozzles as shown in Figure 4.

4. Interrupting air flow in sequence with ejection timing. Here the air is on for only a short time period (burst) when ejection is required.

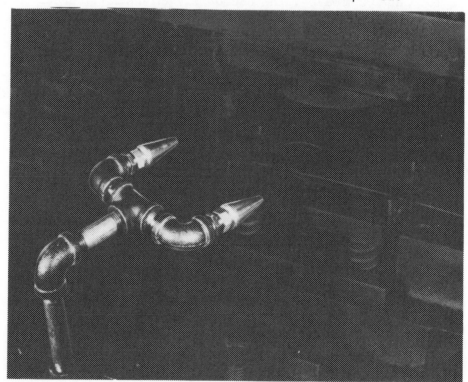

Courtesy of Vortec Corp.

FIGURE 4   High velocity air eductor/diffuser nozzle silencer.

It should be noted that in most of the above measures and especially numbers (3) and (4), less air is usually required at a considerable cost savings. With today's skyrocketing energy costs, one large auto manufacturer estimated a cost of $2,000 to $4,000 per year to continuously operate a 1/4" diameter copper tube ejection nozzle at shop air pressure.

One other factor in air ejection noise control which must always be considered is the sharp rise in noise level as the air stream impinges on a blunt obstacle or flows over a sharp edge. This secondary source of noise has rather efficient radiation properties with noise level increases of 6 to 10 dB common. Air flow over a sharp edge can also generate high noise levels and in some cases discrete whistle-like tones result.

This impingement noise or edge tones can be minimized by avoiding or eliminating sharp edges. One method is to aerodynamically streamline the die surfaces where the air flow is high as illustrated in Figure 5.

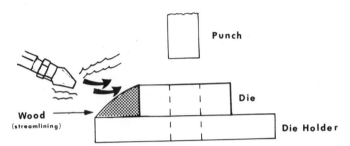

FIGURE 5   Illustration of noise reduction by streamlining air flow over edges.

An alternative to air ejection is mechanical part ejection which is virtually noiseless. There are two basic approaches to mechanical ejection. In the first approach an arm-like mechanism enters the die area, attaches itself to the part (generally with suction cups) and removes the part or scrap. This approach works well on large slowly operating presses where flat sheet metal parts are being formed.

The second approach uses an air actuated cylinder piston to push or eject the part from the die area. The cylinder can be mounted horizontally and the ejection piston impact is automatically sequenced as the ram reaches top dead center or thereabout. This air actuated piston approach does not work well if the parts are very small or flat since the area for piston impact is minimal. A silencer is also required for the air cylinder exhaust which is discussed in a later section.

There is one other measure, which must be considered, that can significantly lower die area noise and that is the die area enclosure as illustrated in Figure 6a. Over the last 5 years, a great deal of design and testing has been applied to the development of these partial enclosures with encouraging results. These enclosures are particularly applicable to automatic presses where continuous operator accessibility and visibility are not usually required. Noise reduction from these partial enclosures is usually substantial but rarely exceeds 6 to 10 dBA. As such, resultant levels at the operator station are typically in the range of 88 to 92 dBA and, at best, marginal with respect to current OSHA regulations. A note of caution: as mentioned, a great deal of development has gone into these enclosures and the design, fabrication and installation is best left to qualified professional or commercial suppliers. In short, a number of suppliers now provide "off the shelf" die area enclosures for a wide variety of common power presses. As such, it is usually less expensive and generally more satisfying to use the "buy rather than make" approach when considering die area enclosures.

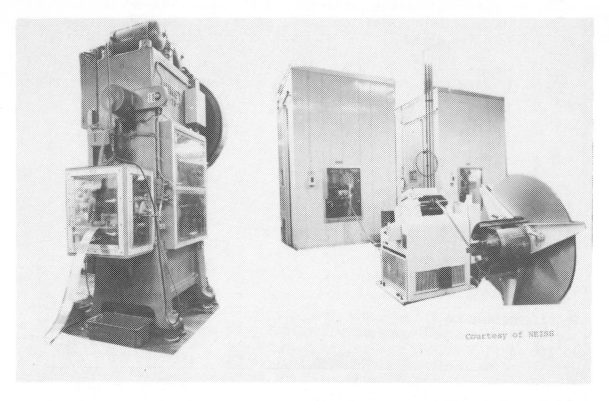

FIGURE 6   Photos of a) die area partial enclosure and b) total enclosure.

Press Equipment Noise: Press equipment noise generally comes from the following sources:

     1.   Motors;

     2.   Clutches;

     3.   Gears, bearings; and

     4.   Pneumatic control exhausts.

The motors, clutch, gear and bearing noise is generally secondary or negligible compared to overall noise. The magnitude of noise levels from these sources can be assessed easily by running the press without dies or workpieces. However, bearing and gear wear or maintenance related items can quickly turn these machines into loud first order noise sources. Fortunately, since they are maintenance related, they tend to get attention quickly or the press becomes real quiet, i.e., it stops.

With respect to pneumatic control exhausts, the noise is also generally secondary, but the large number of controls in a large press room and frequency of discharge result in a nearly continuous noise level. The accumulation of these short bursts of noise is often then of first order magnitude with peak sound levels in the critical 2,000 to 8,000 Hz range.

Treatment of air exhaust noise is often rather easy with satisfying results. Most pneumatic control exhausts have a standard pipe threaded exhaust orifice. A wide variety or porous metal or slotted cylindrical silencers as shown in Figure 7 are commercially available which can be threaded directly into the orifice or readily adapted. These silencers work well but many users have found that they "clog," causing excessive back pressure, and must be cleaned or replaced periodically.

FIGURE 7   Pneumatic control exhaust silencers.

Another, and often better, approach is to adapt a 1/2" or 3/4" diameter (typical) flexible hose to the threaded exhaust orifice and vent the air into an overhead 2" diameter (typical) pipe header running the length of the press room. In this way, numerous exhausts can be vented into a common expansion chamber. The ends of the pipe are left open eliminating back pressure build-up and further maintenance problems.

Structure Noise: Structure noise has its origin during the punching action of the cycle. There are tremendous forces applied and released at impact and breakthrough respectively. These forces are transmitted from the die area to the frame, Pitman, drive shaft, flywheel, etc., and these structures respond with some form of vibratory motion.

Generally, the level of the noise radiated from the press structure is secondary, unless unusual resonant situations exist. More often, high level structure radiated noise originates from relatively lightweight, structural fixtures such as flywheel covers, proximity guards, plumbing conduit, etc. which are loose. The noise character from these sources is more of a metallic "clatter" or "rattle" and can usually be eliminated with some straightforward maintenance with a screwdriver.

For lightweight sheet metal guards, treatment with "paste on" damping material is effective or refabricating the guards from constrained layer damping materials is even better as illustrated in Figure 8.

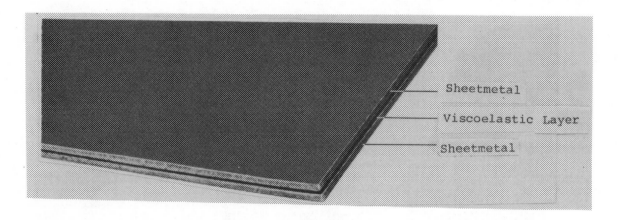

FIGURE 8   Constrained layer damping material.   Courtesy of Antiphon Inc.

One other aspect of press structure related noise that needs mentioning is the isolation of the press itself from the floor. The amount of noise reduction due to isolating the press (isolation mounts) is somewhat nebulous since conflicting reports have ranged from dramatic reductions of 10 dBA to a disappointing 0 dBA at the operators station.

Regardless of the amount of noise reduction, good engineering press installation practice requires high quality isolation mounts for a variety of other reasons and they are strongly recommended.

Material Handling: The high velocity and subsequent impact of metal parts and scrap against tote bins usually is a significant contributor to press room noise. In fact, where large presses (600 to 2,000 ton) and corresponding large parts are being handled, the "metal to metal" clatter can be the dominant source of noise. Consider, for example, the clatter a sheet of steel (soon to be a truck fender) makes when it is placed (thrown) into the press die. Next, comes the modest "crunch" of the forming operation and then again the impact noise as the scrap and fender are thrown into a metal tote bin or stacked. Another example, is the noise made as small parts (several ounces) are blown out of the die onto the side of a metal tote bin and subsequently falling to the bottom impacting other similar parts.

Material handling, like die area noise, is not that easy to control, but listed below are some basic guidelines:

1. Treat areas where parts impact tote bin with rubber or polyurethane impact pads as illustrated in Figure 9.

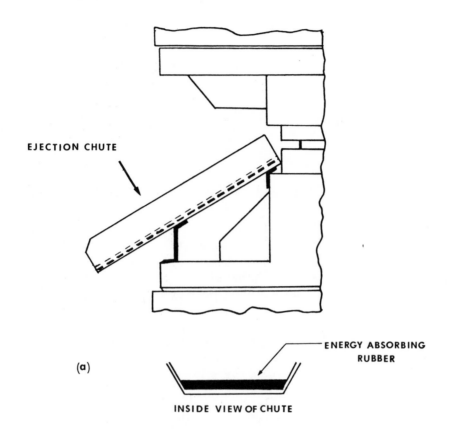

EJECTION CHUTE

ENERGY ABSORBING RUBBER

(a)

INSIDE VIEW OF CHUTE

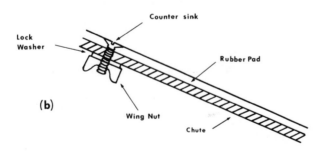

Counter sink

Lock Washer

Rubber Pad

(b)

Wing Nut

Chute

FIGURE 9   Illustration of a) rubber treated ejection chutes and b) typical installation method.

2. Minimize the parts or scrap "drop" distances.

3. Consider enclosing scrap and part tote bins with easy access enclosures as shown conceptually in Figure 10.

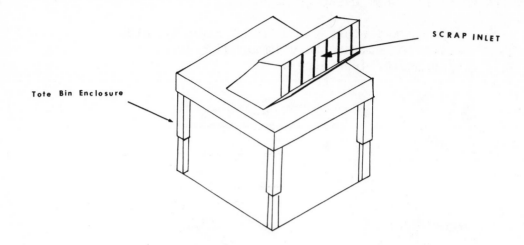

FIGURE 10   Illustration of enclosed scrap or parts tote bin.

<u>Total Press Enclosures:</u> With all previous measures and guidelines applied, press room levels will probably still be marginally excessive with respect to current health and safety criteria. Total press enclosures as shown in Figure 6b, however, do have the potential to bring noise levels down 15 to 25 dBA, as such resultant levels at the operator station are typically in the range of 80 to 85 dBA.

These enclosures have been especially successful when applied to high speed blanking presses where parts and scrap handling is usually simple and continuous operator accessibility and visibility are not required.

With respect to production problems, users have reported little impact on production or production quality. However, it must be emphasized that to assure accessibility for die changes, set-up, parts/scrap removal, and maintainability, additional floor space is usually required. For example, the rear wall of one press enclosure will often seriously restrict accessibility of the die truck or fork lift to the front of an adjacent press or vice versa. A good rule of thumb is: twice as much space for a total enclosure as is typically required for an unenclosed press.

In summary, to assure a successful total press enclosure installation:

1.   Detailed studies of the problems associated with parts and scrap removal and die area accessibility must be conducted; and

2.   Based on these studies, sufficient floor space must be provided.

If the available floor space is tight or marginal, it is strongly recommended that total press enclosures be considered infeasible and the program abandoned.

Here again, as in partial die enclosures, the design, fabrication, and installation of total press enclosures should be left to commercial suppliers with qualified backgrounds of experience.

Plant Layout: Some reduction in employee noise exposure can be accomplished by following common sense guidelines in press room layout.

As a first consideration, isolate (when possible) typically noisy equipment from quieter operations. In particular, it may be possible to "wall off" high speed blanking operations thus reducing exposure to adjacent manual press operators and inspectors. The wall may be of simple "stud" construction, acoustical panels or flexible acoustical curtains. Also, if machines used infrequently are placed adjacent to noisy equipment, the extra spacing will provide noise reduction to the operators in the line beyond.

Obviously, the application of these press room layout guidelines are easiest to implement and more effective when expansions or new machine installations are in the planning stage.

SUMMARY

In summary, press room noise control must be considered as one of the most challenging technical disciplines required of today's metal working engineer. There are emphatically no simple "gimmicks."

The guidelines presented herein, if applied, will in most cases, provide measureable noise level reduction and in many cases, significantly lower operator exposure. It must, however, be emphasized that a coordinated effort between the tool designer, manufacturing engineer, set-up and the maintenance department is essential for effective results.

REFERENCES:

1. Robert C. Locke, "Automatic Strip Feed Press Noise and Its Reduction", Reduction of Machinery Noise Seminar, Dec. 10-12, 1975, Ray M. Herrick Laboratories, Purdue University.

2. Robin H. Alfredson, "Noise Source Identification and Control of Noise in Punch Presses", Reduction of Machinery Noise Seminar, Dec. 10-12, 1975, Ray Herrick Laboratories, Purdue University.

3. O. A. Shiniashin, "Sources and Control of Noise in Punch Presses", Office of Research and Development, Environmental Protection Agency.

4. R. D. Bruce, "Noise Control for Punch Presses", 21st Annual Technical Conference, American Metal Stamping Association, New York, N.Y.

5. J. A. Daggerhart and Elliot Berger, "An Evaluation of Mufflers to Reduce Punch Press Air Exhaust Noise", Noise Control Engineering, May-June, 1975.

6. C. Allen and R. Ison, "A Practical Approach to Punch Press Quieting", Noise Control Engineering, Vol. 3, No. 1, July, 1974, p. 18-20.

7. O. A. Shinaishin, "Impact Induced Industrial Noise", Noise Control Engineering, Vol. 2, No. 1, 1974, p. 30-36.

8. N. D. Stewart, J. R. Bailey, and J. A. Daggerhart, "Study of Parameters Influencing Punch Press Noise", Noise Control Engineering, Vol. 5, No. 2, Oct., 1975, p. 80-86.

Reprinted from Metal Stamping, August, 1978

How to Diagnose an Ailing Press

As is the case with ailing people, the first step in diagnosing a weary press is a thorough "physical." It isn't difficult and the symptoms it discloses can reveal all manner of both obvious and hidden problems. It also points out the direction in which the "cure" may lie.

By Robert R. Campbell, President, Robert E. Campbell, Inc.

This article is written in the realization that in the typical stamping plant, production is paramount. The presses must keep running. But even in the busiest plant, with limited press capacity, the work can be scheduled in such a way that a given press can be taken out of service long enough for a thorough examination.

If the press passes its physical, very little has been lost. If the ex-

. . . if the press passes its physical, very little has been lost

amination turns up symptoms indicating an early breakdown, production can be scheduled in such a way as to allow you to plan for any necessary repairs. This is preferable to an unplanned stoppage because of a broken crankshaft, sheared connection, cracked frame, etc., and the ensuant frantic calls for help.

All presses should be given a physical from time to time. Obviously, priority should be given to equipment that is giving problems. If a die that has been in and out of a press a dozen times suddenly begins to produce bad parts, the press, rather than the tooling, may be the problem. The bolster may have been bowed; the crankshaft may have bent under an overload.

The first step in the examination process is to go over the press frame inch-by-inch, looking for cracks and breaks. This is a lot easier if the press has been cleaned up first. It's hard to find a crack under a 15-year accumulation of dirt and grime.

If the press was bought used (or if it's a press you're thinking of buying) watch for welded or brazed repairs. There's nothing wrong with a brazed

or welded frame, provided that the job was properly done. But cracks in the frame are never an isolated symptom. The forces that cracked the frame may have bent the crankshaft or severely worn the pitmans, bearings and gears.

This point merits further comment. A press is a self-destructive device. It's a tribute to the press builders that they last as long as they do. (Bliss recently turned up a number of its presses, still in service, that were built in 1890.)

But a press is also a balanced collection of separate parts. The motor drives a gear which drives a shaft which, through a connection, moves a ram which is confined in ways on the frame. Anything that upsets this balance, whether it be worn bushings, a sprung frame or scored gibs has to affect all of the other components.

. . . usually a major failure has been building up over a long period of time

Thus, if a crankshaft bearing is worn out-of-round because it wasn't properly lubricated, it isn't just the crank that is affected. Accurate alignment is lost; the gears wear, the ram cocks and all of a sudden you can't make good parts.

For this reason, if you call in a press repair specialist because of a broken crankshaft, he will probably merely glance at the crankshaft but he will spend a lot of time looking at other things. He will want to know *why* the crankshaft broke and what other damage has been done *before* it broke.

Sometimes, because of a severe and sudden overload, a single com-

ponent will break. This is rare. Usually a major failure has been building up over a long period of time — which is exactly *why* presses should be carefully examined from time to time.

How often? My own recommendation is at least twice a year.

In checking the frame, look at all the bearing surfaces, especially the ram ways, for signs of scoring or gouges or other signs of wear. If the ways on one side are shiny from sliding contact and the ways on the other side are covered with old grease, the ram is not coming down square and you can expect future trouble.

Similarly with the gibs. With a strong light, any scoring or wear on the gibs stands out like a red flare on a dark road. Scored gibs can indicate lack of lubrication, improper adjustment, out-of-parellelism with the slide, off-center loading or even loose mounting bolts. Caught early enough, you may be able to take corrective action.

Now take a good hard look at the connection. In older presses with a ball screw design, the ball screw itself was a planned weak link. The theory was that the ball screw would fail first, avoiding overload conditions in the frame or crown.

In practice this didn't always work but the connection will usually start showing signs of sickness early on.

Next step is to check parellelism between the bed and ram. It isn't difficult. Simply lower the slide and zero an indicator at any given point on the upper die holder. Moving the indicator along the lower die holder surface will show any deviation from parallelism.

User standards vary, but generally deviation shouldn't be more than .001 inch per foot. If it's greater than

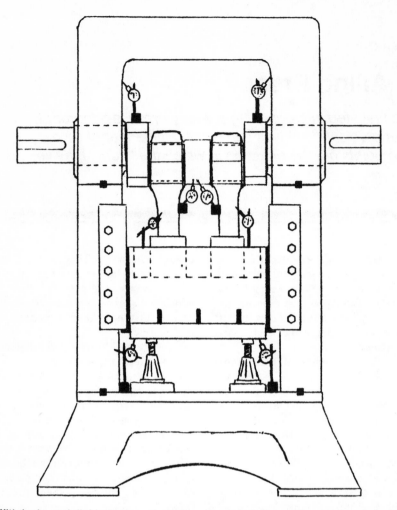

With jacks and dial indicators check the clearance between the crank to frame bearings, connections to the crankshaft, ram to connections and bed to ram. On a two-point press, as shown here, the readings should be consistent from one side to another.

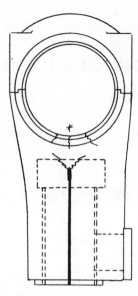

X-es and jagged lines show probable failure areas in the connection. Damage is usually the result of overloading but it is often due to careless die setting; for instance failure to tighten the lock plugs after ram adjustment. This can strip threads and crack the connection.

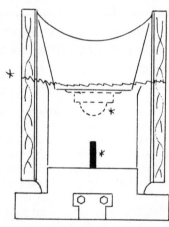

This is the ram of a single-throw OBI. Check for cracks across the back of the ram, in the ball seat pocket and knockout holes. A major cause of failure is improper location of the die; usually this amounts to placing it too far forward for operator convenience in loading.

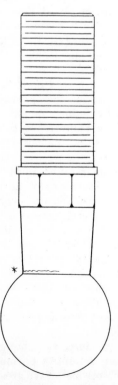

The ball screw will usually crack in the area around the neck but its a good idea to check the threads for burrs or cracks. A sudden overload can break the ball screw without further damage to the rest of the press but this is unusual.

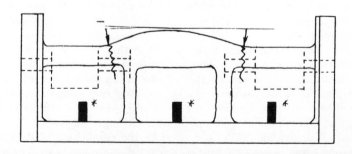

This is the cast iron ram of a two-point press. Check around the knuckle seats and the knockout pin holes for cracks. Jagged lines indicate areas most likely to crack. Probable cause of failure is off center loading. If you are running two 50-ton and one 100-ton jobs in a 200-ton press you are creating an off-balance condition.

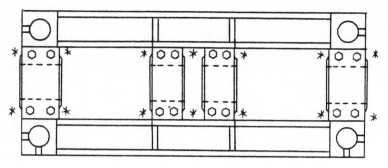

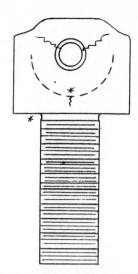

This is the crown of a two-point fabricated press. X-es show the points where weld failure is most likely to occur. A press with minor cracks in these areas can be operated, if necessary, with great care but repairs should be scheduled as soon as possible.

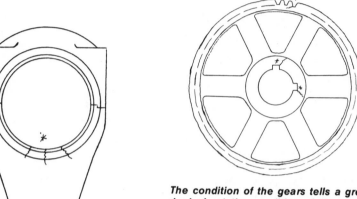

Lower connection which mates with the upper connection wrist pin. Check around the pinhole, the knuckle seat, the radius and also the threads for cracks. Faulty lubrication of the upper connection wrist pin will often translate itself to damage here.

The condition of the gears tells a great deal about the condition of the press. Check the teeth at the root line for cracks. Gears will often fail at the keyways. Keys should be tight, otherwise a hammering condition exists which can do great damage. When gears break or show excessive wear it's usually because of problems elsewhere along the line such as a bent crankshaft or badly worn gibs.

Upper connection wrist pin showing areas most apt to crack. One area is as apt to crack as another despite size difference. The wrist pin hole should be carefully checked. If wear is detected early, an in-plant repair can be made. Common cause of failure is faulty lubrication and the severe burden on the lower pin aggravates it.

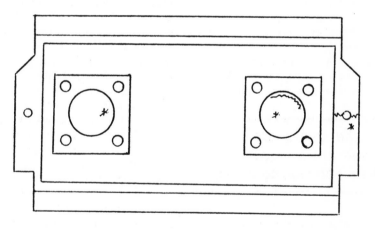

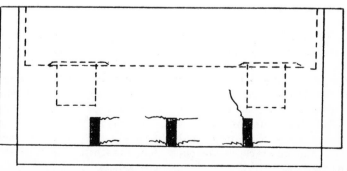

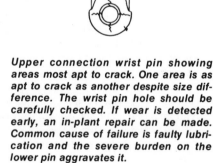

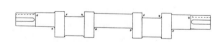

X-es show areas most apt to crack on a two-point crankshaft. Check radii along the stress axis for cracks and (important) check keyways in the gear ends for twist. Since they are very strong, crankshafts often bend without fracturing. When this happens, all components are forced to operate abnormally. Usually cause is overloading; more often through diesetter error than operator carelessness. (Double feed.)

This is a steel ram and, while it won't readily crack, it can become distorted. Check the bottom of the lower connection seat, all weld seams, and the counterbalance rod holes for cracks. A ram of this type, overloaded beyond its deflection range, will become sprung and can be used only a considerable risk to your tooling. Improper setting of knockout pins is a major cause of damage to the ram.

this the frame may be sprung and you are now risking damage to your tooling.

Another check is simply to place a straight-edge across the bed. If daylight shows under the straight-edge, the bed is bowed) probably from an overload. Check the bed front-to-back *and* side-to-side. Bowing can take place in either direction.

It's a good idea to check the Tee slots and bolt holes at this point. Hammered down Tee-slots and worn out bolt hole threads aren't particularly harmful in themselves but they can be costing you a great many die-setter-dollars.

The most difficult part of the examination is checking the bearings. To do this, using dial indicators (see the drawings with this article for details) and jacks, raise the ram and measure the play in the bearings. No flat tolerance can be given here, lacking knowledge of the age and type of press, but it should not be excessive and it should be consistent across the throws.

The final step in the static inspection (press shut down) is to look at the gears. They can tell a lot. If the press is in perfect alignment, the gears do not wear. Uneven or excessive wear on the gears or broken teeth mean only one thing. The gears are trying desperately to compensate for *something wrong elsewhere* in the chain of events that is a press.

On many presses the crankshaft isn't visible without disassembly. If

---

. . . the gears are trying desperately to compensate for something wrong

---

it is accessible, it should be checked for cracks in the area around the throw(s).

Dynamic examination — power on — can pick up many symptoms not otherwise detectable. The first step should be to cycle the machine and to actuate all the controls in every mode provided by the builder; from Inch to Full Automatic. Many press users use only one mode in normal operations and never check the optional modes.

A good sense of hearing is useful in dynamic testing. Old timers used to say that they could tell if a press was running right by simply listening,

and they weren't far wrong.

By this I don't simply mean that they could pick up the sound of the ram hitting too hard or bottoming which is unmistakable. They could also tell a great deal from the gear noise. Bent or broken shafts, worn bearings or bushings, worn or damaged gears or pinions, loose pinion or gear keys, insufficient lubrication and even excess backlash as a result of overloading produce characteristic sounds — noises, actually — and the gears make a very effective stethoscope.

. . . slippage can wear out clutch parts faster than you can replace them

---

Dynamic testing, incidentally, should take place with counterbalance pressure off. It is literally true that you can mask many press defects by setting the counterbalance pressure way up. In doing this you are deluding yourself and setting the stage for future shock — shock in the form of heavy repair costs.

Listening to the clutch can also be informative. One common cause of clutch slippage is insufficient air pressure. Clutch slippage can wear out clutch parts faster than you can replace them. Conversely, too much air can damage clutch components. From my observations, clutch air pressure is improperly set on as many as 25% of presses.

The clutch in good condition doesn't groan or squeal. The in and out action is firm and positive and the sound is crisp.

At this point you should stand back and watch the press cycling. The gears shouldn't wobble. If the press is of tie-rod design, look at the joints between the uprights, bed and crown. They should not open up under load. Nor should the crankshaft bushings gap on either the up or downstroke.

If you see grease or lubricant oozing from any joint or bushing at a particular point on the stroke, then the press is "working." In effect it is twisting itself to try and compensate for something: very often a bent crankshaft.

Your last step should be to put your hand on the ram and to hold it there for a few cycles. I have done this repeatedly and without looking

at the press I have diagnosed a dozen different ailments ranging from egg-shaped bearings to worn keyways.

It isn't mumbo jumbo or witch doctory. It's completely logical. Remember that the press is a collection of components and the press action is a series of events. If the main bearings wear, the crank clearance changes and every other component has to adjust.

All the various clearances change and become nonconcentric. At the bottom of the stroke and the top of the stroke, all the clearances are taken up at the same time. This can be felt, even on a new press in perfect condition. As the clearances widen, there becomes a sort of dwell where all the clearances come together and one can almost feel the weary press gathering itself together to make the big effort.

The increasing shock can be felt and, with some experience, identified with the components involved. In a really worn press it is little short of terrifying — even if you don't own the press.

In the drawings with this article press components are shown with the probable fail areas marked and causes identified. I have examined literally thousands of damaged presses and inquired into the cause of the damage. I have come to some conclusions that may be of interest.

First, all press failures happen on the night shift. We all know that this can't possibly be true but that's the story I usually hear and I'm sure that you do too.

Seriously, the prime cause of press damage is probably diesetter error. You will find very few diesetters who will acknowledge this (or who will, in fact, acknowledge any error of any kind).

This reflects two things: the critical nature of diesetting where a minor error can produce a monumental overload, and the generally poor training of diesetters.

The second major cause of press damage is inadequate lubrication and it doesn't lag very far behind the primary cause. Reason: lubrication is entrusted to the low man on the totem pole in the maintenance department and he is neither properly instructed nor supervised.

Proof of this is that new presses with fail-safe automatic lube systems have far lower failure rates due to wear than older equipment with manual greasing.

The third cause of major press damage is operator error; almost invariably because he double fed or triple fed a secondary operation. I recently completely rebuilt a 250-ton press in which the operator, hired that day, had loaded *fifteen* parts with removing any.

The press withstood the pressure as long as it could. When it finally let go, it almost literally exploded. One can hardly blame the press builder.

As I mentioned earlier, the physical exam can tell you a lot about your equipment.

Some of the illnesses disclosed can be treated in-plant but most require the services of a specialist.

I do not suggest, should you find a heart murmur or a blocked artery or a sprained ankle, that you shut the equipment down. Usually, if you baby it, you can continue to operate it — carefully — until you can schedule downtime for repair by a specialist at a time designated in advance to your mutual benefit.

Just don't wait too long.   ○

A prescription for sick presses

Reprinted from Precision Metal, July, 1980

The first step in diagnosing a sluggish press is a thorough "physical." It is not difficult to do and can reveal where the cure lies.

Just as people need a physical check-up periodically to maintain their health, so do forging presses, hammers and upsetters, and stamping and punching presses as well.

Robert Campbell, president, Robert Campbell, Inc., Lansing, Michigan, contends that, even in the busiest plant, with limited press capacity, diagnostic work can be scheduled so that a press can be taken out of action long enough for a meaningful examination. The Campbell company's entire business is repairing presses — it does not sell new presses — so the statements Bob Campbell and his engineers and technicians make are worth listening to.

If a press passes an examination, little production time has been lost. However, if the physical exam reveals symptoms that may lead to an early breakdown, production can be arranged so that necessary repairs can be made before a serious breakdown occurs. This kind of maintenance preparedness is preferable to unplanned stoppages in the midst of a production schedule.

Campbell has been involved in repairing, rebuilding and recycling presses for 31 years. He has told of his method of maintenance at numerous seminars, such as those sponsored by the Society of Manufacturing Engineers and the American Metal Stamping Association. On such occasions, the audiences have been large, and the question/answer sessions have been long. The Campbell company employs approximately 30 people.

"Press users get into a rut of taking these workhouses for granted," Campbell said. "They lose sight of the fact that a press is basically a self-destructive device. It is a tribute to press builders that they last as long as they do. Age is not a big factor in their ability to function efficiently, if they are systematically cared for and minor repairs are made before major breakdowns occur. Presses should be given a thorough examination at least twice a year."

A major failure will generally progress over a long period of time and affect more than one major part, because a press is a balanced collection of separate parts. The motor drives a gear, which in turn drives a shaft, through a connection, which moves a ram that is confined in ways on the frame. Whatever upsets this balance, whether it is worn bushings, a sprung frame, or scored gibs, will affect all the other components.

The diagnostic process

The first procedure in examining a tired press is a visual inspection of the frame. Examine all bearing surfaces, especially the ram ways, for signs of scoring, gouges or other signs of wear. If the ways on one side are covered with dirty, old grease, the ram is not coming down square, and future problems can be expected.

Watch also for cracks and breaks that may develop from the inside out. If welding or brazing is necessary, make sure that you get them performed by an expert in the field.

Next, look closely at the gibs. Any scoring or wear will stand out like a red flare on a dark road, when a strong light is used during the inspec-

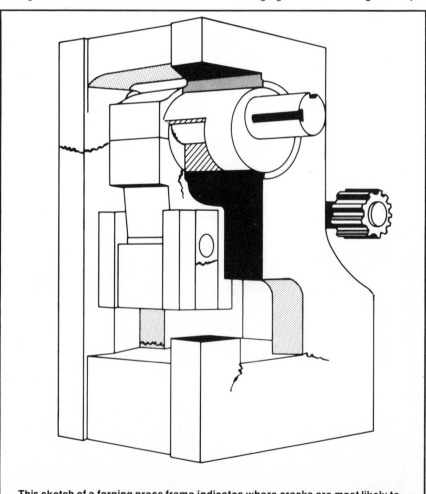

This sketch of a forging press frame indicates where cracks are most likely to develop from overloading and off-center loading. The cutaway view shows an eccentric shaft which might break down at the right-hand bearing carrier.

tion. Scored gibs often indicate one of five hidden problems: lack of lubrication; improper adjustment; out-of-parallelism with the slide; off-center loading; or loose mounting bolts. Once the problem is spotted, early corrective action can be taken.

In older presses utilizing a ball screw design, it is important to inspect the area around the neck, because this is usually where cracks develop. It is also a good idea to inspect the threads for burrs or cracks.

"The ball screw was designed to be a weak link," Campbell explained, "and in theory it should fail first, thus avoiding overload conditions in the frame or crown. In practice, however, this does not always happen. Users should begin looking immediately for the causes of ball screw damage and correct them before risking damage to the rest of the press."

To avoid possible damage to the press frame or to your tooling, Campbell recommends that the user determine if any deviation in parallelism has occurred between the bed and the ram.

"It is not difficult to do this," Campbell says. "Simply lower the slide and zero an indicator at any given point on the upper die holder. Now move the indicator along the lower die holder surface. Any deviation from parallelism will be immediately noticeable. Generally, deviation should not exceed 0.001 in./ft.

Anything more than this usually is a sign of a sprung frame.

"Next, inspect the bed front-to-back and side-to-side by placing a straight-edge across the bed. If daylight shows under the straight-edge, the bed has probably been bowed as a result of overloading."

The most difficult part of the visual portion of the press physical examination is the inspection of the bearings, according to Campbell. To do this, you will need a dial indicator and a jack. Raise the ram and check the clearance between the crank-to-frame bearings, the connections to the crankshaft, the ram to the connections and the bed to the ram. The readings should be consistent from one side to the other as you move around the press."

The final step in the static examination process is a visual inspection of the gears. If the press is perfectly aligned, no wear will be evident on the gears. Uneven or excessively worn gears, however, mean that there is something wrong elsewhere in the press that is forcing the gears to try to compensate.

Gears will often fail at the keyways.

Keys should be tight, or else a hammering condition will occur. When gears break or show excessive wear, check for a bent crankshaft or badly worn gibs.

Use your ears and hands

Put the press under power, and have an experienced member of your maintenance staff complete the final stages of the press examination. Ascertain that the area has been posted or roped off to prevent unauthorized personnel from entering the test area.

"Put the machine under power, and rely on your sense of touch to tell you where excessive wear might be developing that you cannot see," said Campbell. "Place your hand on the frame, preferably with the press under load, and concentrate. You will feel certain vibrations that will indicate misalignment or excessive wear.

"All the various clearances change and become nonconcentric. At the bottom and the top of the stroke, all the clearances are taken up at the same time. This can be felt, even on a new press in perfect condition. As the clearances widen, there is a sort of dwell where all the clearances come

A 700-ton forging press is in the process of being rebuilt at Campbell, Inc.

This 3-in. upsetter machine has been completely rebuilt.

together, and you can almost feel the weary press gathering itself together to make the big effort. Moving around the frame and noting the vibration, you will be better able to zero in on the problem area."

Next turn off the safety device and place your hand on top of the ram and hold it there for a few cycles. Campbell says, "I have done this repeatedly and without looking at the press have diagnosed a dozen different ailments ranging from egg-shaped bearings to worn keyways. Pay close attention to vibration which you feel, and with a little experience you will be able to do the same thing. However, be careful — never perform this test without the knowledge of the operator, or without posting the area off limits to other personnel."

Then use your sense of hearing to further diagnose potential press ailments. Take the counterbalance off, as this can mask many press defects. Then listen to the gears and the clutch.

Bent or broken shafts, worn bearings or bushings, worn or damaged gears or pinions, loose pinion or gear keys, insufficient lubrication and

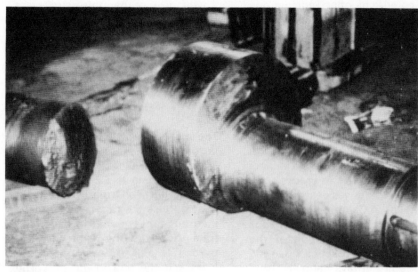

A 2500-ton forging press crankshaft that has been broken and taken to the Campbell plant in Lansing, Michigan for repairs.

even backlash as a result of overloading produce characteristic sounds and noises, which make the gears very effective stethoscopes.

Three major causes of press damage that can be controlled within the plant are diesetter error, operator error and inadequate lubrication. Campbell states that he has examined thousands of damaged presses, and he believes the prime cause of press damage is attributable to diesetter error.

Die casting machinery maintenance

Equipment supplied by a vendor is only as effective as the organization that stands behind it.

A sophisticated approach of planned maintenance is available from companies like Lester Engineering Company, Cleveland, a manufacturer of die casting machines and injection molding machines. Lester's plan provides scheduled maintenance by highly qualified factory trained service representatives at a very low cost.

The Planned Maintenance Program (PMP) is contracted for a minimum of four quarterly one-week visits per year at approximately 80% of the prevailing "on call" service rate.

A list of duties that will be performed during the one-week period follows: hydraulic and electrical inspection of equipment with a submitted report; general troubleshooting and physical maintenance work; operator training in die casting procedures; a review of the maintenance program in existence in the particular plant; and general die casting inspection. In addition to a detailed diagnosis of each machine condition, the technician conducts eight hours of illustrated lectures of training and updating in hydraulics and electrical systems.

The PMP program serves two functions: 1) Every quarter it quantifies the quality and extent of maintenance performed over the preceding three months. These periodic inspections provide an indication to management of the effort being exerted to protect their highly invested die cast installation. 2) The frequent interrelationship of the visiting technicians with employees on a periodic basis serves to increase the knowledge of the maintenance people

A planned program of inspection at quarterly intervals also provides management with an expert opinion on prevailing operational procedures.

Presented at the SME Understanding Presses And Press Operations Clinic, December, 1979

Preventive Maintenance For Mechanical Punch Presses

By George Petruck
President
Power Gems, Inc.

In most stamping plants today, we find two types of maintenance.
They both start with the letter "P", but there the similarity
ends. One is "PANIC" maintenance, which is helter-skelter,
mass confusion with no control. The other type is "PREVENTIVE"
maintenance, which is organized with a well-trained staff and
controlled. This latter type is what we will be covering.

The three topics to be covered are:

> I.   Objectives of a preventive maintenance
>   program.
> II.   How to set up a preventive maintenance
>   program.
> III.   How to implement the program.

I. <u>OBJECTIVES</u>:
1. Increase reliability of machines.
2. Reduce downtime to a minimum.
3. Decrease maintenance costs.
4. Increase productivity (profit).

Reliability of machines

The primary objective of a preventive maintenance program is to
insure that the machinery is functional and in good working
order; or, to insure maximum productivity.

Reduce downtime to a minimum

Downtime has a direct effect on production, Productivity and
profit decrease in direct proportion to the increase in downtime. As an example of the harm in downtime, losing 150,000
parts results in a loss of $7,500 (.o5/part x 3 weeks). This
is exclusive of the cost of repair (see attached chart).

Decrease maintenance costs

With the installation of a good preventive maintenance program costs can be reduced. It is cheaper to repair a punch press than it is to rebuild one, i.e., a warm bearing is an indication of future problems; if ignored, the problem could be compounded.

Productivity

If objectives 1-3 are met, it is a foregone conclusion that there will be an increase in production, with a corresponding increase in profits. Many articles are written about how we are not keeping up with foreign competition. One reason is that the competition has, and uses, preventive maintenance as a workable tool.

II. HOW TO START A PREVENTIVE MAINTENANCE PROGRAM

1. Job description.
2. Maintenance check sheets and work orders.
3. Day-to-day and long-range planning.
4. Priorities.
5. Spare parts and standardization.
6. Budget.

Job description

In starting a preventive maintenance program, we need people who are trained and qualified to do the work, supervised by a leader who is both competent and efficient, with a background and working knowledge of machinery. It is not only important that he be an expert in electronics, lube systems, machinery, et cetera, but he must surround himself with people who have these capabilities. One of the most important members of his crew should be an experienced millwright: A person who knows bearing clearances, slide parallelism, disassembly and assembly procedures, and who has an inquisitive mind. A competent millwright can save countless dollars in being able to diagnose a problem and present information to the foreman, who can then decide to pull a press for minor repair or to have it rebuilt. Going back to the warm bearing, we've seen that, for an example, using a lapping compound, such as "timesaver", on a bearing saved three weeks of downtime.

Such determinations are made from experience or, unfortunately, by learning the hard way. The second crew member should be a competent electrician. Even with the advent of solid state controls, which have self-checking systems indicating where the problem is, there are thousands of control systems in use today that require the services of a good electrician who can read and diagnose an electrical schematic. Finally, the crew should have a person knowledgeable in lubrication systems. Lubrication is an area that is overlooked in many plants today. In the past, most lube systems consisted of zirk fittings to individual lube points on a press. These were later changed first to fittings to a central point, then to automatic lube; and finally to re-circulating systems. Knowing when and how to lubricate is just as important as being able to do electrical and millwright work. It is important that prerequisites with regard to experience and job knowledge be established in order to insure the efficiency and competency of the maintenance crew in dealing with problems that arise on a day-to-day basis.

Check sheets and work orders

As problems arise and repairs are required, all too often work orders are issued verbally. There are then no written records to establish maintenance costs. To that end, in order to keep a log of work done, a work order sheet should be issued to list the problems, the scope of work to be done, the time started, the time finished, and any recommendation for future work. Attached is a copy of an example work sheet, and the following will examine some of the items that are important.

Periodic Planning

Again, we can refer to the check sheet and establish the planning that goes into a preventive maintenance program. For example, simple things like draining the water from the air lines can greatly increase the clutch and brake life. Or, making sure the oil or grease reservoir has ample lubricants could save a bearing. Long range programs can be much more involved. Checking the clutch and brake for wear could be a monthly program, as could checking the slide-to-bed parallelism. You could establish a yearly program where presses are taken out of production and given a complete check-up.

Priorities

If you have three presses down, (1) an automatic press with progressive dies, (2) a blanking press, and (3) a press doing secondary operations, what would your priorities be? I am sure that most of you will agree that the automatic should come first. However, setting priorities is a problem that should be discussed with the production people since this concerns their operations.

Spare parts and standardization

The next item that we will discuss is having an adquate supply of spare parts available. How many days are lost to downtime due to a $10.00 item? With most press companies, delivery of shelf items is normally one week; and if we refer to the attached chart on profit loss due to downtime, even one week could cost over $2,500 @ .05/part. Therefore, parts subject to wear should be kept on the shelf. (e.g. linings, springs, packings, seals, relays, et cetera.) Most press companies indicate those parts in their service manuals which are recommended spare part items; but, unfortunately, too many people fail to read their manuals or ignore the advice of the experts. Closely allied with spare parts is standardization and/or interchangeability of parts. For those of you who work

for companies whose presses are all the same, this does not
present a problem. Even though such uniformity is the exception rather than the rule, at least investigate the possibility
of standardization of repair parts.

III. <u>IMPLEMENTATION</u>

Now that we know what our objectives are, and how to set up a
preventive maintenance program, the final step is how do we
implement the plan? The most important personnel who are
needed along with a competent technical staff is management.
If your management is concerned with reducing downtime and
increasing profit, then the problem is minimal. Remember,
maintenance is usually the low man on the totem pole; but,
if we can upgrade the maintenance department, then we can for-
get <u>PANIC MAINTENANCE</u> and concentrate our efforts of
<u>PREVENTIVE MAINTENANCE</u>.

MAINTENANCE CHECK SHEET

MACHINE NO. _____ TONNAGE _____ TYPE _____

STROKE_____ S.P.M. _____

SHUT HEIGHT _____ S/N _____ MAKE _____

INSPECTED BY				
DATE				
CONDITION				
I. CLUTCH & BRAKE				
A. Mechanical				
B. Air Friction				
Linings				
Springs				
Packings				
Seals				
Pressure Sw.				
II. DRIVE				
Crankshaft				
Main Gears				
Pinions				
Main Brgs.				
Conn. Brgs.				
Flywheel Brgs.				
Keyways				
Flywheel Brake				
III. SLIDE				
Gibs				
Liners				
Adj. Screw				
Adj. Nut				
Pitman				
Wrist Pin				
Ball & Seat				

continued - -

136

lV. BED				
Cushions				
Ways (Ext)				
V. C'T BAL. CYL.				
Packings				
Piston & Rod				
Pressure Sw.				
VI. TIE ROD & NUTS				
VII. LUBRICATION				
Oil Lines				
Fittings				
Supply				
VII. ELECTRICAL				
Panel				
Relays				
Wiring				
Palm Buttons				
"V" Belts				
Motor-Main				
Motor-Slide Aj.				
Motor-Lube				
Foot Pedal				
IX. AUXILIARY EQUIP.				
Air Blow-off				
Pull Backs				
Barrier Guards				
K.O.				
Feeds				
Straightners				
Reels				
Scrap Choppers				
Conveyors				

PROFIT LOSS DUE TO DOWNTIME

PARTS/MIN	PARTS/WK*	.01	.02	.05	.10
20	40,000	$ 400	$ 800	$ 2,000	$ 4,000
25	50,000	500	1,000	2,500	5,000
30	60,000	600	1,200	3,000	6,000
40	80,000	800	1,600	4,000	8,000
50	100,000	1,000	2,000	5,000	10,000
100	200,000	2,000	4,000	10,000	20,000
200	400,000	$4,000	$8,000	$20,000	40,000

* Based on 50 minute hours - 1 shift (8 Hours) - 1 Week
  Example: 25 Parts/Min x 50 Min Hr. x 8 Hrs. x 5 days
               = 50,000 Parts/Week

Since most press accidents involve crankshafts:

Assume delivery time 8 weeks (10 weeks downtime)

Cost of Crank	$ 5,000.00
Cost of Bearings	1,000.00
Cost of Labor	2,500.00
Cost of Clutch Parts	1,000.00
COST OF PARTS AND LABOR	9,500.00

ASSUME press capable of producing 25 parts/min.
25 Parts/Min = 50,000 parts/week x 3 weeks = 150,000 parts
25 Parts/Min = 50,000 parts/week x 18 weeks = 900,000 parts

PRODUCTION LOSS	PART PRICE	SUBTOTAL	REPAIR COST	TOTAL LOSS IN DOLLARS
900,000	.01	$ 9,000	$ 9,500	$ 18,500
900,000	.02	18,000	9,500	27,500
900,000	.05	45,000	9,500	54,500
900,000	.10	90,000	9,500	99,500

Repair-Rebuild-Replace: Factors To Consider For Justification

By George Petruck
President
Power Gems, Inc.

INTRODUCTION

With the ever-increasing cost of doing business today, especially with high interest rates and increased material and labor costs, one of the problems facing management today is whether or not to repair, rebuild, or replace existing machinery. Our topic will center its attention on metalforming machinery, and specifically punch presses.

In the metalcutting field, numerous advancements have been accomplished in the last 30 years: Tape Controls, Automatic Tool Changers, Higher Feeds and Speeds, etc., to enhance the state-of-the-art. However, in the metalforming end of it, we seem to be dragging our feet. There have been some improvements, such as Hydraulic Overloads, moving bolsters, higher speeds, wet-type clutches, solid state electrics, improved lubrication systems and a few other improvements. These changes have been mostly on Straight-Side presses and very few changes have been made on OBI and OBG type presses. OBI presses are the most abused machine tool on the scene today because:

(1) They are constantly overloaded.
(2) Very little, if any preventive maintenance is done.
(3) In many instances, they are operated by unskilled help.

Because of the above factors, and additional ones that we will see, we are going to discuss the three "R's".

(A) Age of the Press

With over 200,000 presses in use today, the majority of them are over 30 years old. Most of these presses have:

1. Mechanical clutches - pin type, jaw, rolling key, and some with direct drives
2. Zirk lubrication fittings
3. Spring or no counterbalance cylinders
4. Out-dated electrical systems
5. Line-shafts
6. Little or no guarding

The first question that we ask ourselves is: Is it economical to repair or rebuild a 30-year-old machine?

(B) <u>Tonnage</u>

Presses under 75 tons could be border line when it comes to rebuilding. (Figure 1) The cost of rebuilding a 60-ton OBI is approximately 85% of the cost of a new one. We will discuss the repair and rebuilding costs later.

COST % - REBUILDING VS. REPLACING

TONNAGE	%	NEW
35	100%	$18,500
45	90%	$22,000
60	85%	$27,000
75	75%	$32,500
110	65%	$53,000
150	60%	$66,000

FIGURE 1

The second question is: Should we spend $23,500 to rebuild a 30-year-old 60-ton press?

(C)   Press Utilization

The next factor to consider is: How many hours a shift is the press being used? If you were to walk through your plant three times per shift, you might be amazed at how many presses are idle. If you are not getting full utilization from your presses, this may be the opportunity to upgrade your equipment.

(D)   Productivity of Floor Space

Using a nominal figure of $8.00/sq. foot as the cost of floor space, and assuming that a press occupies 60 feet, the cost of the space comes to $480.00/month/machine. If we replace a machine running 60 S.P.M. with a press running 120 S.P.M., we have increased our productivity, but our cost remains the same. We can take this one step further. By upgrading our press from a 60-ton to a 75 or 110-ton press, with variable speed, we may be able to reduce the number of presses from three to one, thereby reducing our floor space costs, and yet increasing our productivity.

(E)   Loss of Productivity

Another factor to consider is the amount of parts that we will lose when we repair or rebuild a press. If we have a press that is capable of producing 25 parts/minutes and it will take three weeks to repair the press, we have lost 150,000 parts. Following this line, if we rebuild the same machine, our down time stretches to 18 weeks, with a loss of 900,000 parts. (Figure 2)

LOSS OF PRODUCTIVITY

**25 PARTS/MIN.   50,000 PARTS/WK
(BASED ON 50 MIN.
HOUR)**

25 × 50 × 8 × 5 = 50,000

**3 WKS DOWNTIME = 150,000 PARTS
(REPAIR)**

**18 WKS DOWNTIME = 900,000 PARTS
(REBUILD)**

FIGURE 2

LOSS OF PROFIT

150,000 PARTS AT $.02/PART = $3,000

900,000 PARTS AT $.02/PART = $18,000

FIGURE 3

Loss of Profit

As you can see from Figure 3, at $.02/part profit, our loss would be $3,000.00 plus the cost of repairing the press, and $18,000.00 loss of profit when we rebuild the same press. We will point this out later when we discuss the costs involved.

(F)   Costs

    1.   Repair
    2.   Rebuild
    3.   Upgrade to OSHA
    4.   Replace (Used)
    5.   Replace (New)

    1.   Cost to Repair

As our example (Figure 4) we will use a 60-ton press with a mechanical clutch and a 110-ton OBI, with an air clutch and brake. The cost to repair a 60-ton OBI with a new crankshaft, new main and connection bearings, clutch parts, etc. is approximately $8,900.00. The cost for the 110-ton is approximately $11,900.00.

FIGURE 4

Cost to Repair

	60 TON	110 TON
Crankshaft	$ 5,000.00	$ 6,000.00
Bearings	$ 1,000.00	$ 1,400.00
Clutch Parts	$ 900.00	$ 1,500.00
Labor	$ 2,000.00	$ 3,000.00
TOTAL	$ 8,900.00	$11,900.00

With the press out of production for three weeks, our loss of profit is $11,900.00 for the 60-ton, and $14,900.00 for the 110-ton press.

2. <u>Cost to Rebuild</u> (Figures 5 and 6)

If we decide to rebuild the press and upgrade it, the costs involved would double. The procedure that most rebuilders follow is: Disassemble, clean and inspect the complete press. New parts are then obtained (either from the press manufacturer, or the rebuilder will make them from dwgs of damaged parts). The press is remachined (Bed, Slide, Gibs, etc., new electrical Lube System and an Air Clutch would be installed). The cost of the rebuild comes to $23,500.00. We add this cost to the loss of productivity (or profit) and the real cost comes to $41,500.00. ($18,000.00 loss of profit plus $23,500.00 cost of rebuild). Figure 6 shows the cost to rebuild a 110-ton press: $33,000.00 plus loss of profit or $51,000.00.

COST TO REBUILD & UPDATE

60 TON OBI

A. DISASSEMBLE, CLEAN, INSPECT	$2,500
B. NEW PARTS	$8,000
C. CLUTCH CONVERSION	$5,000
D. NEW ELECTRICS	$3,500
E. RE-MACHINE GIBS, BOLSTER, SLIDE, ETC.	$2,000
F. REASSEMBLE	$2,500
	$23,500

FIGURE 5

COST TO REBUILD & UPDATE

110 TON OBI

A. DISASSEMBLE, CLEAN, INSPECT	$5,000
B. NEW PARTS	$11,500
C. NEW ELECTRICS	$4,500
D. RE-MACHINE GIBS, BOLSTER, SLIDE, ETC.	$6,000
E. REASSEMBLE	$6,000
	$33,000

FIGURE 6

3. Cost to Replace (Used)

If we decide that the downtime of 18 weeks is too long, we may decide to replace our press with a used machine. Figure 7 reflects the cost of a 60-ton press and amount required to upgrade it to OSHA ($17,000.00). We are assuming that all that is required is electrics and misc. clutch and brake parts. This would also be true with a used 110-ton press. (Figure 8)

COST TO REPLACE

60 TON OBI – USED

COST	**$12,000**
ELECTRICS	**$3,000**
MICS. CLUTCH PARTS	**$2,000**
	$17,000

FIGURE 7

COST TO REPLACE

110 TON OBI – USED

COST	**$33,000**
ELECTRICS	**$3,500**
	$36,500

FIGURE 8

144

4. Cost to Replace (New)

The last avenue we can explore is to buy a new press. If we rebuild a 60 or 110-ton press, we still have a 30-year-old machine. The approximate cost of a new machine is:

$$60\text{-ton.} \ldots \ldots \ldots \$30,000.00$$
$$110\text{-ton.} \ldots \ldots \ldots \$53,000.00$$

There are other factors that should be considered in making our decision to repair, rebuild or replace. They are:

1. Trade-in value

Even though we have an old 60 or 110-ton press, they have a certain trade-in or resale value. The amount that you can expect will vary depending on how you dispose of it.

2. Freight & Erection

Taking the machine out and shipping it to a rebuilder is a factor that cannot be overlooked, as well as putting the press back in service. (Electrical hook-up, piping, etc.).

Figure 9 is a summary of the costs involved, and can be a guide for determining the course of action that you can take.

FIGURE 9

Summary	60 TON	110 TON
Repair	$ 8,900.00	$11,900.00
Rebuild	$23,500.00	$32,000.00
Replace - Used	$17,000.00	$36,500.00
Replace - New	$30,000.00	$53,000.00

We have not tried to come to any conclusions as to whether it is more economical to repair, rebuild or replace a punch press. All we have tried to accomplish is to point out some of the factors involved. The final decision is yours.

CHAPTER 4

CONTROL OF THE STAMPING PROCESS

Prepared for *Understanding Presses And Press Operations*

Measuring Loads On Mechanical Presses

By Donald F. Wilhelm
President, HELM Instrument Co., Inc.

This article provides information to help in understanding the use of strain/stress measurement devices and electronic instruments to measure **and** analyze deflections in metal-stamping press structures.

For many years it was largely an art, and very little science, to set up and adjust tooling in metal-stamping presses. It is now possible to use electronic technology to solve some perplexing press problems. Techniques have been developed to conveniently measure stress/strain levels in press structures and to diagnose press and tooling illnesses. Many dangerous or non-productive load levels may be detected and corrected. Press-load analysis is providing much-needed information to efficiency-concious metal-stamping engineers and management people.

There are many reasons why it is important to measure loads on metal-working machines.

1. Protect Press

   It is necessary to assure that forces required to make parts do not exceed the capacity of the machine structure.

2. Protect Tooling

   With the advent of sophisticated and expensive tooling materials, every effort should be made to assure only the minimum force be used to make the part. This will not only increase tool life but minimize press component wear as well.

3. Select the proper size press

   If the tonnage requirement of a particular tool is known, it aids in selecting the proper size press. This is most useful in those operations where a variety of tools and frequent changes are made.

4. Determine Press Condition

   One of the more important advantages of being able to measure loads is the ability to determine the condition of the press and recognize press problems. Press-analysis can be used to develop preventative maintenance programs thereby avoiding unexpected downtime.

5. Adjust tools properly

   Having the ability to accurately measure loads enables the operator to adjust the stamping process to assure proper tool and press operation which results in better part quality, less wear on tool and press components and will provide the opportunity to increase speed for higher productivity.

6. Energy Savings

The forming of parts in the metalworking process is the result of using the kinetic energy available from the flywheel. Consuming this energy in the working cycle reduces the flywheel speed hence its energy. This energy can only be replaced via the electric motor. Anytime a part is hit harder than is necessary it results in a waste of electrical energy. Depending on load and number of strokes per minute, this can be significant.

MEASURING FORCES THROUGHOUT THE PRESS STROKE

Presses are rated in three categories:

1. STRUCTURAL CAPACITY

This rating determines the stiffness of the slide and is based on predetermined deflection standards on uniformly-distributed loads. Therefore, concentrated forces not uniformly distributed can impose severe overloads on press members even though the total force is below the press capacity. Measuring these loads on the uprights of a straight side press will show front-to-back as well as side-to-side off-center loading

2. TORQUE CAPACITY

This rating affects the design of rotating drive members. The higher up in the stroke the force is applied, the greater the force on these parts due to the change in crank angle. Low forces high in the stroke can impose severe overloading on crankshaft, bearings, clutch, etc.

The measurement of either the angular position of the crankshaft or the linear position of the ram, while the force on the press is being measured, is an excellent means of knowing just where the maximum forces occur.

3. ENERGY CAPACITY

This rating determines the size of the flywheel and motor, based upon the distance through which the rated load can be applied, and the frequency of force application. Moderate forces applied through too long a distance, or at too rapid a stroke rate, can stall the press and/or overheat the motor.

"SNAP-THROUGH"

'Snap-Through' forces occur on blanking operations when the part breaks through the die suddenly releasing the compressive energy so subjecting the press members to forces in the reverse direction.

Manufacturers generally rate presses at approximately 20 percent tension capacity. If these 'snap-through' forces are excessive, severe damage can occur even through the compressive load is well below the press rating.

For these reasons it is important that force-measuring instruments be capable of reading these peak reverse forces. The bidirectional and dynamic measurement ability of the strain-gage is ideal for this.

METHODS OF LOAD MEASUREMENT

There are two major methods of measuring load in mechanical presses.

1. Incorporate the measuring elements as integral parts of the tooling (fig.1). This is particularly useful in multi-station tooling where individual station forces can be separated. Until recently, the cost of such measuring elements has been quite high because in most cases the sensors are custom designed to be an integral part of the tool and must remain with the tool. Many times the increased production, increased tool life, and quality of the finished part more than justify the higher sensor costs. New developments in sensor technology and electronics has provided the industry with a multitude of standard and semi-standard transducers at greatly reduced cost. Examples of such elements are shown in fig.2.

2. Make the press a measuring element by recognizing the deflections in the press structure as a function of the load in the die area. For example, deflection or strain measuring elements can be attached to the press structure and calibrated by applying a known force in the work area. This force may be applied statically by means of hydraulic jacks, or preferably dynamically through the use of calibrated load cells.

   Once the strain level has been measured in a particular location of the press structure the tools can be changed. Forces developed by other tools can be accurately measured without the need for further recalibration.

WHERE ARE THE FORCES MEASURED?

In the case of stamping presses the forces required to make the part are distributed throughout the press structure. In gap-frame (fig.3) or OBI presses, there is naturally the compressive force on the connections as well as the tension forces on the front part of the frame and a compressive force on the back of the frame. The neutral axis through the frame is the result of the ratio of frame thickness on the front to the frame thickness on the back. Generally the reinforcing on the front tends to shift this axis toward the front and since the strain literally rotates about this axis the strain levels are considerably greater to the rear of the structure.

In case of straight-side solid-frame presses, obviously the frame is in tension and the forces on the connections are compressive. On tie-rod presses the 'pre-shrink' of the rods puts the uprights under compression. When work is being done the tension force on the tie-rods relaxes the compressive force on the uprights in proportion to the tension force. In this location it is very easy to mount measuring elements and they are generally out of the way of any moving parts. However, should the force on the press exceed the pre-shrink on the tie-rods, the crown will lift meaning there is no longer a compressive force on the uprights. Without adequate indicating instruments it is impossible to recognize this and a 'peak' force measurement would be in error. However since rods are generally pre-shrunk to a level of about 200 per cent of the press capacity, the uprights are an excellent location for monitoring force.

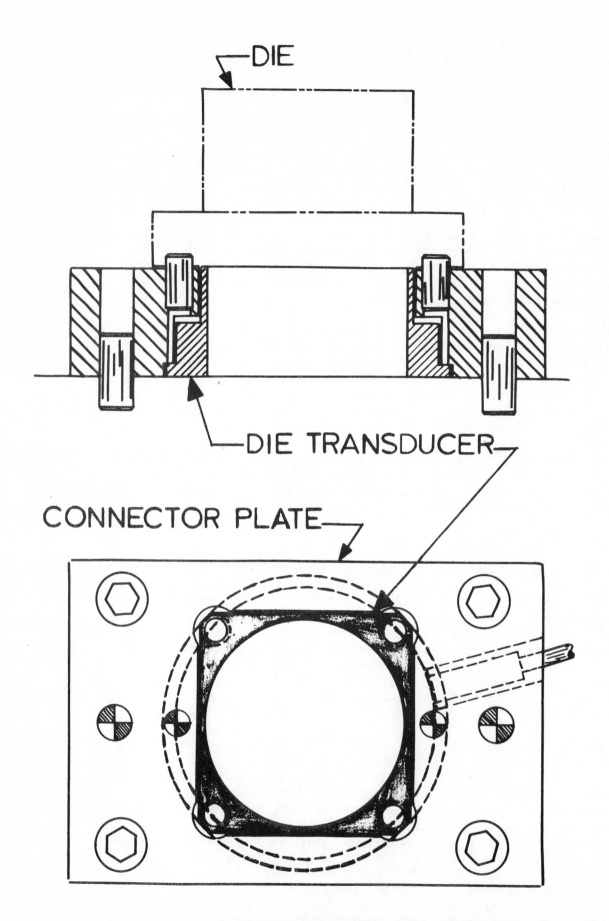

SCORE DIE TRANSDUCER

FIGURE 1

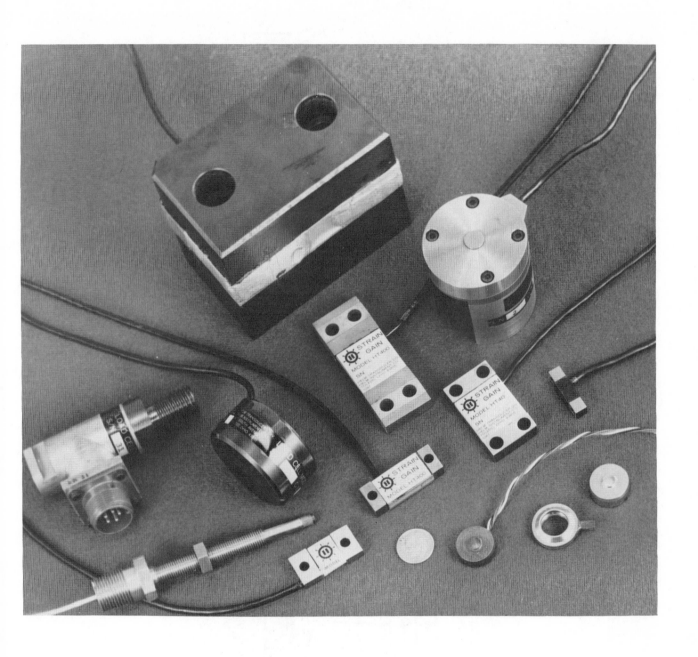

SENSORS FOR TOOLING

FIGURE 2

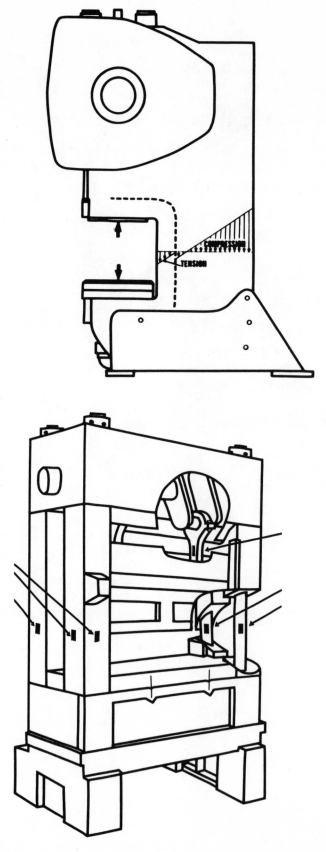

MEASUREMENT POINTS ON O.B.I. AND STRAIGHTSIDE PRESSES

FIGURE 3

ADVANTAGES OF LOAD MEASUREMENT

1. Strain-gage transducers, which are bidirectional, measure either tension or compression as well as reading static or dynamic loads and provide an indication of loss of preshrinking rods to assure uniform loading of uprights.

2. Die-set-up time can be greatly reduced and jobs can be set to pre-viously-established optimum conditions of loading and load dis-tribution. Off-center loading can easily be detected.

3. Early detection of change in stock thickness, die lubrication or metal characteristic help maintain quality.

4. Measurement of tool wear helps in scheduling maintenance on an optimum bases. By incorporating the appropriate readout instru-ments it is possible to capture and observe the peak forces, follow the varying force through the full press stroke, indicate where in the press stroke the peak force occurs and measure the time shift in peak force from one part of the press structure to another.

5. Accurate parts counting can also be done by allowing a counter to operate only after the press has reached a pre-determined force.

FORCE-MEASUREMENT SYSTEMS

It is very desirable that the forces generated in a metalforming operation are displayed as accurately as possible and in a form con-venient to the user. There has been a lack of confidence in many earlier systems due to inaccuracies caused by ambient temperature changes, energizing voltage variations, as well as long-term no-load signal drift. Solid-state electronics have now enabled systems to be produced where such errors are of neglibible magnitude.

Measurement of Strain

Measurement of strain requires the ability to recognize dimension-al changes in the order of millionths of an inch. For example, struc-tures subjected to forces in the order of 10,000lb/sq. in. represents strain levels of about 400 micro-inches.

When force is applied to a press structure, the change in dimen-sion that occurs is a function of the cross sectional area of the structure and it's modulus of elasticity or ability to resist change in dimension. We know, for example, that the cast irons used in press structures have a modulus of about 20×10^6 PSI. Steels normally found in press structures have a modulus of about 30×10^6 PSI.(fig.4)

Measurement of strain requires the ability to recognize dimension-al changes in the order of millionths of an inch. As an example:

A column of cast iron 12" x 12", with a 6" tie-rod-clearance hole through the structure, has an effective cross sectional area of 12" x 12" = 144 sq. in. less the 6" dia hole cross section of 28.27 sq. in. = 115.7 sq. in. If a load of 100 tons (or 200,000 pounds) is applied, the change in dimension of the column will be:

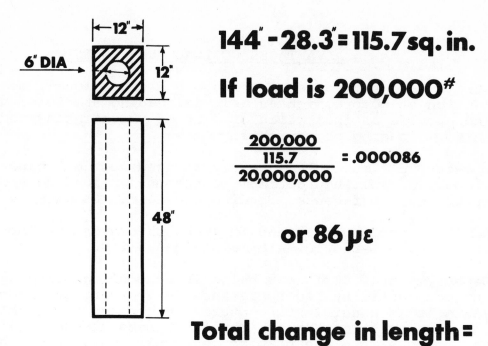

$144'' - 28.3'' = 115.7$ sq. in.

If load is 200,000#

$$\frac{\frac{200,000}{115.7}}{20,000,000} = .000086$$

or 86 $\mu\varepsilon$

Total change in length =

.000086 × 48 = .0041

FIGURE 4

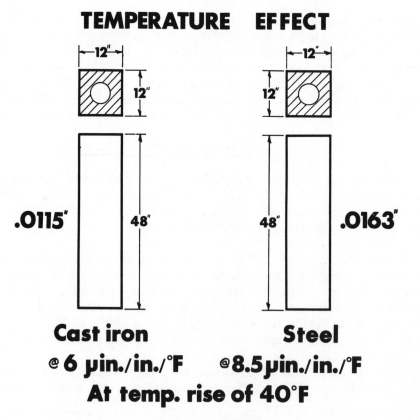

TEMPERATURE EFFECT

.0115″   **.0163″**

Cast iron   **Steel**
@ 6 μin./in./°F   **@8.5μin./in./°F**
At temp. rise of 40°F

FIGURE 5

$$\frac{\frac{200,000}{115.7}}{20 \times 10^6} = .000086"/inch$$

If the column is 48" long the total change in dimension will be:

$$.000086 \times 48 = .0041"$$

This change illustrates the stringent requirements and specifications for the measuring element and the associated electronics in the hostile environment of a stamping plant. This .0041" change becomes significant if demensional change of the column is providing the load data. A load measurement instrument can be calibrated to directly relate the change-in-demension to the change-in-load.

Temperature Effect

An often overlooked factor in strain measurement is the effect of temperature on structures. When one considers that all materials change dimension with change in temperature and consider the two most common materials found in press structures, cast iron and steel, change significantly. We find the coefficient of thermal expansion to be very important, about 6.5×10^6 in/in/°F for iron and 8.5×10^6 in/in/°F for steel (fig 5). Refering to the previous example of dimensional-change vs. load-change we had a change in length of the 48" tall iron column of about .004 for a load change of 200,000 lbs. If we consider the same column with a temperature coefficient of 6.5×10^6 in/in/°F and a modest change in temperature of only 40°F we then have a change in column length of about .0125" or 3 times the dimensional change due to a 100 ton load. In the case of steel, with a higher coefficient the change would be over .016". This clearly illustrates the importance of compensating the load measurement systems for this temperature effect in order to obtain a valid measurement of load. There are electronic means available today to accomplish this and are commonly referred to as "automatic zero balance" circuits. Care must be exercized in selecting an instrument with automatic zero circuits since they must be able to respond at the proper rate to assure not effecting the true load level plus they must be continuous and automatic in operation.

ELECTRICAL-RESISTANCE STRAIN-GAUGES

Numerous methods have been used to measure stress/strain levels. The most popular has remained the strain-gage. The reason for this popularity has been; ease of installation relative insensitivity to shock, ability to measure rapidly changing force-levels, ease of obtaining repeatable measurements, high quality with minimum cost, dependable data when used with quality instruments. . . and much more.

In 1856 professor W. Thompson (Lord Kelvin) described numerous experiments performed in investigating the electro-dynamic properties of metals. Perhaps one of the most significant findings was the observation that the electrical resistance of certain wires varied with the tension to which they were subjected. From this early work the electrical-resistance strain-gage has been intensively developed.

If a wire is bonded to a large structure it becomes essentially a part of the structure. As the structure is strained the wire will experience some deformation due to the applied strain.

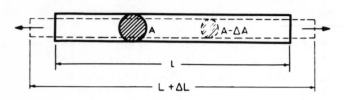

ELECTRICAL RESISTANCE : $R_1 \approx \frac{L}{A}$

$$\Delta R = R_2 - R_1 \approx \frac{L+\Delta L}{A-\Delta A} - \frac{L}{A}$$

FIGURE 6

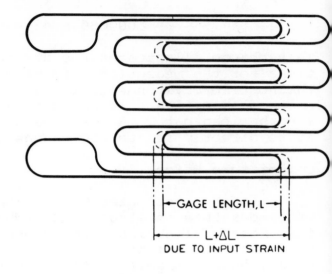

GAGE LENGTH, L

L+ΔL

DUE TO INPUT STRAIN

FIGURE 7

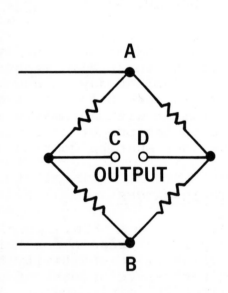

WHEATSTONE BRIDGE

FIGURE 8

BOLT-ON TRANSDUCER ON WELD PADS

FIGURE 9

The electrical resistance of wire is a function of its cross-sectional area and its length. If the wire is elongated (fig.6) then as the length increases the cross sectional area is reduced and the resistance increases. Conversely, if the wire is compressed the length decreases, the area increases and its resistance to current flow decreases. By forming a grid (fig.7) and effectively increasing the length and resistance, the measuring accuracy is substantially increased.

The strain-gage elements are in the form of a simple electrical Wheatstone Bridge. (fig.8) Each arm contains one strain-gauge element. The bridge is energized by a very stable D.C. voltage fed across terminals AB. Any initial electrical out-of-balance is eliminated by a balance control potentiometer. The strain-gage bridge circuit is applied to the structure to be measured.

When a press force is generated the structure deforms and the resulting strain unbalances the bridge. A minute voltage, proportional to this force, appears at the output terminals CD. This output is fed via shielded cables (to reduce spurious signals) to the input of the measurement system.

The use of strain-gages provides a very reliable strain measuring technique with a high degree of immunity from external disturbances such as large and rapid changes in temperature, acoustic noise, mechanical vibration and shock.

Attaching a sensitive, delicate strain-gage directly to a press raises problems which make this method unattractive. The objections are:

1. The best adhesives, providing the necessary long-term stability, require curing at elevated temperatures (up to 400 deg F). Heating a large press structure by radiant heat is difficult and uncertainties exist as to the actual temperature of the adhesive. The alternative is to use a room-temperature-curing adhesive which is undesirable for a permanent installation.

2. Extreme care must be exercised in applying the gages since the operation of the gage depends on its intimate contact with the structure. Surfaces must be very clean and free from all oils and contaminates. It is difficult, outside a labortory, to maintain these standards of cleanliness.

3. The attachment of gages should not interfere with the production schedule of the machine. Using high-temperature adhesives could mean that the press would be idle for at least 24 hours.

4. Any damage to the strain gage, subsequent to the press calibration means that a press is again idle while new gages are attached and the press is re-calibrated.

BOLT-ON AND WELD-ON TRANSDUCERS

Objections to strain-gages are removed by bonding the strain-gages under laboratory conditions to an intermediate structure which can then be bolted, or welded to the machine structure under test (fig.9).

This method also allows for mechanical amplification and further increases the measurement accuracy. The bonding cement can be cured in laboratory conditions under a carefully-controlled time-temperature program.

A further advantage in the use of 'Bolt-On' and 'Weld-On' strain-gage transducers is that the individual units are interchangeable. Thus, once a machine is calibrated, transducers can be interchanged without re-calibration.

CALIBRATION

The process of adjusting the force-measurement-instrument to agree with the actual load on the press is referred to as 'calibration'.

The calibration procedure varies with different applications. For example, there are two common and proven ways to calibrate a mechanical or hydraulic press, either with a 'load-cell' (dynamic calibration), or with a hydraulic jack (static calibration). With either method the purpose is to apply a known force to the press structure. The dynamic calibration technique involves the use of a mass of metal with strain-gages connected to it; this equipment is commonly referred to as a 'load-cell' (fig.10) and is actually a large strain-gage transducer which gives a calibrated output for a given force.

The load-cell must be placed on the press bed in a suitable location (fig.11), prefereably centered, or evenly spaced if more than one load cell is used. It is important to begin the calibration with a small momentary force on the press to protect the load-cells from a possible damaging overload. The shut height would then be adjusted down after each single stroke of the press until the desired load is placed on the press structure as indicated by the load-cell output. For the known force, which can be now placed on the press structure, the strain-gage transducers bolted to the structure provide a given output which is displayed on the force-measurement meter.

The sensitivity of the measurement instrument is then adjusted so that the indicated meter-force reading corresponds to the known calibrated load-cell reading.

Providing a record is made of the sensitivity setting of the measurement instrument for a particular press, the instrument can be used with any number of presses simply by choosing the correct setting for each press.

Determining Press Condition

How often do we install very expensive and close tolerance tools in presses without knowing if the press is capable of allowing the tool to perform to its optimum? Various press conditions can be determined through the use load measurement. For example:

Parallelism of Slide to bolster

The traditional method of checking parallelism is to sweep the open-

PRESS CALIBRATION LOAD CELLS

FIGURE 10

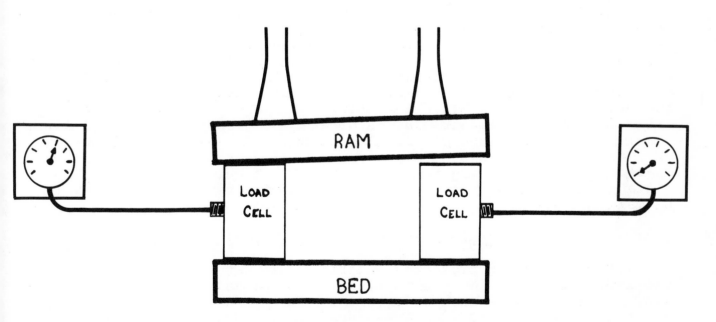

LOAD CELLS ON PRESS BED

FIGURE 11

ing from slide to bolster with a dial indicator with the press not running. This does not necessarily take into account the difference in bearing clearances. In those cases where blocks or jacks are used to support the slide, care must be exercised to as not to introduce bending and deflections that would mask the true parallel condition. Static checking of parallelism does not take into account those deflections that are a result of mass in motion creating forces found only when the press is running.

The following is a suggested method of dynamically measuring parallelism.

By striking on two or more calibrated load cells in the bed area and using shims to develop equal loading on each cell then measuring the thickness of the shims and knowing the distance between cells a very accurate measure of parallelism can be made. If by this process the press frame is calibrated then future checking of paralellism can be made by striking on ground parallels in lieu of calibrated load cells, and reading load distribution on an instrument.

In the case of OBI or gap frame presses it is important to be able to recognize the angular deflections that occur in this style press. Despite the fact that there is no industry standard for angular deflections, most press manufacturers design around the value of .0015" per inch of throat depth at rated capacity. Considering a nominal throat depth of 20" this could mean an angular deflection of as much as .030"! Ask yourself, "Can the tool accept this much misalignment?" There is also the misconception that guide pins and bushings can compensate for this misalignment. One must realize the size of the pin and bushing could easily consume all of the working area of the press leaving no space to make parts.

By using the same technique of measuring parallelism, the amount of angular deflection in gap frame presses can be determined and most times steps can be taken to compensate for this deflection in the setting of the tool. A calibrated load cell (with shims) is placed in the die area in line with the centerline of the crankshaft and and slide lowered until sufficient force is exerted on the cell when the press is running. Then the load-cell can be moved in the die area. If the load measurement is greater or less than the beginning load, it means that non-parallelism exists. To calculate the amount of deflection that exists, shims can be added or subtracted from the load-cell. By measuring the thickness of the shims and knowing the distance the cell was moved, the angular deflection of the press may be calculated.

Cardiagrams for Presses & Tools

All metalworking processes exhibit a "signature" of characteristic force time relationship that, if measured and recorded, can be very useful in analyzing performance. This is somewhat analagous to the record made in studying human heart performance. A previously recorded chart of a normal heart operation is compared to a chart made when a problem is suspected. This comparison can be invaluable in the diagnosis. The same principle can be used in the metalworking process. In the case of stamping presses, it is also important to know not only the load, but also where in the stroke the load is developed. An often overlooked specification of the press builder is the tonnage capacity at some maximum distance off bottom dead center. This is illustrated in fig.12.

Several methods are used to recognize the slide position with respect to load. A sensor or 'transducer', fitted to the press to measure the slide position. This transducer provides an electrical output that is recorded on a chart recorder. On the same chart, having the same time base, is recorded the load thus providing an easily read record of the load vs slide relationship.

PRESS-LOAD ANALYSIS

A typical analysis of a force curve is illustrated in fig. 13. This chart is a recording of the loads on the left and right side of a press containing a double die. The staggered tooling is forming aluminum spin-top closures from scroll cut stock.

At first glance it is obvious that one side of the press is subjected to a higher load level than the other side. This peak load is caused by the coining operation. The load could be more evenly distributed by mearly shifting the tooling on the low-load side or by shimming the tooling.

In this case higher load was the result of unparallism of the slide with respect to the bolster. A slight shimming of the die shoe on the low-load size, plus a change of shut-height, resulted in a more even loading. These minor changes resulted in an overall reduction in load. An additional benefit was a better part was made with less wear and tear on the tooling.

It is well worth mentioning that the time lost and expenses of the load analysis was more than repaid by increased production. The speed of the operation was raised from 208 strokes per minute to 280 strokes per minute. This 26% increase, plus better parts, more than justify the tests.

THE LOAD LEVELS THROUGHOUT THE STROKE ARE AS FOLLOWS:

A) This is the blanking operation. In this illustration it is a minor force relative to the force of the coining operation. However, in some other load studies this was a major press-damaging operation. Not so much from the cutting-of-material standpoint but damage resulted when the punch broke through and "snap-through" reverse forces were generated.

B) This is the draw-and-forming portion of the stroke. In most cases this is a relatively gentle applied force. Press damage rarely develops here.

C) This is the coining operation. As mentioned before, this is the prime load in this application. Lets assume the center of each chart represents the maximum usable tonnage of the press. We then have a dangerous condition existing. . .and we might still not be making a good part. Load measurement instruments graphically illustrate how much you are overloading. The instruments then display the effect of adjusting the tooling.

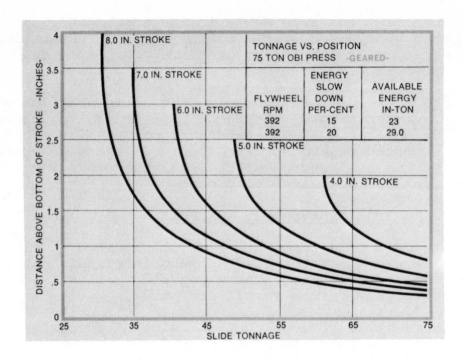

TONNAGE CAPACITY OFF BOTTOM DEAD CENTER

FIGURE 12

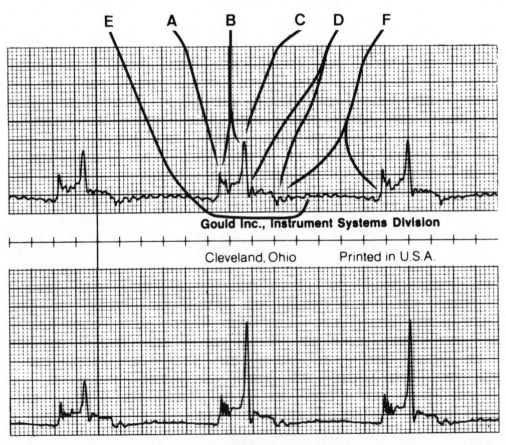

FORCE CURVE ANALYSIS

FIGURE 13

D) Point D represents the stripper operation. In this case all is well. However, in other applications we have seen excessive subtractive stripper loads. An excess negative load here could indicate potential damage to the part being formed. Or, it could indicate why and when parts are getting hung up on the tooling. Obviously erratic stripping action is a potential hazard to the tooling and the press.

E) This is top dead center.

F) This area represents the slide action. Theoretically this should be a flat and smooth section of the recording. . .obviously in this case it is not. This indicates slide chatter. It is possible to see a binding slide here. However, notice that one side of the press has a fairly clean and smooth action and the other side is coarse. . .but not excessive. This slide-binding load could be intensified when the press is doing heavy work. . .such as coining. . .and could be a significant factor in press overload. A minor adjustment of gib-bearing clearance or re-lubrication might quickly cure this problem.

CONCLUSION

Reliable data obtained from press-force monitors in metalforming operations provide the user with valuable information for solving traditional problems associated with tool design, effective press operation, and preventive maintenance. The tool designer can avoid press and tool-overload conditions.

The production engineer can use his presses to obtain optimum operating conditions necessary on such expensive capital equipment. Tool setting time is no longer guesswork. In addition, 'down time' can be considerably reduced. Probably most important of all is the greatly improved ability to maintain uniform quality in produced parts. The safety aspects of the press operation as with other types of machine tools is always a topic constantly in the thoughts of safety officers and the relevant government departments. It is interesting to note that the USA Occupational Safety and health Administrative Law requires that 'employers shall operate presses within the tonnage and attachment-weight ratings specified by the manufacturer 'and' that all dies shall be stamped with the tonnage and stroke requirements or have these characteristics recorded if they are readily available to the die setter.

Stamping presses equipped with force measurement and monitoring instruments will provide the user with the information necessary to comply with such regulations. When the press is calibrated, various die-sets can be used and the actual tonnage required to make parts to specification measured.

In addition to merely measuring the force-levels in a machine, it is desirable to also control the machine. This brings us to load monitoring, with provisions for stopping a press if load levels grow excessive. With the ever-increasing cost of exotic tooling, protection of that tooling becomes a major management concern.

Prepared for *Understanding Presses And Press Operations*

Hydraulic Overload System

By F.E. Heiberger
Manager, Electronic Development
Danly Machine Corporation

A mechanical stamping press has the capability of producing tremendous work due to the energy available from the flywheel of the machine. A standard mechanical press is designed to produce full tonnage with 15% slow-down of its flywheel. This 15% reduction in speed represents a decrease of 28% of the energy available in the flywheel. As an example a 1000 ton press when producing full load still has almost 2600 tons of energy remaining in the flywheel.

If mechanical interference is encountered in the die area of the press due to some foreign material, the press will attempt to complete a power stroke producing tremendous stresses on the gears and links of the machine, as well as the die. In order to limit the tonnage that a press can produce and protect the mechanism, a hydraulic overload system has been utilized that relieves the mechanical interference thereby limiting the peak tonnage that the machine will encounter.

The Hydraulic Overload System (Figure 1) consists of:

1) Press Lubrication Pump & Motor providing oil throughout the press to strategic points i.e. gibs, bearings, etc.

2) An Oil Pressure Gage is located between the 25 micron purolator filter and the high pressure pump to indicate the superchange pressure being supplied to the pump. This pressure should be a minimum of 15 PSI to prevent cavitation for satisfactory high pressure pump operation with press lubrication oil. A pressure of less than 15 PSI could indicate a faulty press lubrication pump, a broken or loose connection, or the elements in the purolator filter are clogged and should be changed.

3) Hydraulic Pump driven by a separate electric motor with multiple positive pistons which are so arranged that each piston can be tapped off separately and in effect become a separate circuit.

One piston is used to maintain pressure on the slide overload circuit. The remaining excess pistons are connected internally in the pump and are used as a rapid means to quickly prefill the connection cylinders after an overload.

The pump is pressure fed with the press lubrication oil which is supercharged to a minimum pressure of 15 PSI to insure an ample supply of oil at the high pressure pump inlet. Before entering the pump, the oil is passed through a 25 micron purolator filter to maintain cleanliness in the overload system.

4) A Hydraulic Slide Pressure Gage is located on the side of the hydraulic control valve manifold on the pumping unit. This gage indicates the pressure delivered by the pump to each connection.

5) A 2-Position Spring Offset Solenoid Controlled 4-Way Valve which directs the excess pistons in the pump to recharge the large hydraulic cylinders after an overload.

6) A Hydraulic Cylinder built into each slide connection. The travel of this cylinder is one inch. The area X the precharge pressure X 1.33 determines the connection tonnage.

7) Slide Overload Valve located in a pilot bore in each hydraulic overload cylinder. All overload relief valves are supplied with oil thru a common supply line. Since the valve spools are spring seated, the oil passes thru the check valve located within the spool charging and maintaining the connection cylinders to system pressure.

Light weight valve components which reduce the inertia effect and large relief ports are provided to instantaneously relieve any overload pressures when they are attained.

The side opposite the mounting pilot diameter is provided with
an oil inlet from the power unit. Two drains are provided on the valve.
The first drain consists of four large annular ports on the forward end
of the valve. Oil from the overload cylinder passes thru these ports
and into an oil compartment which is drained into the main sump. A
second drain in the valve is provided with a 3/8" pipe threaded outlet
connection. This port vents to atmosphere. The main valve spool is
spring loaded to insure that the valve closes after an overload.

8) The Valve Manifold is so ported and arranged to result in the minimum
amount of tubing and connections. It also contains integral checks to
isolate pressure lines from tank lines.

9) Master Relief Valve is a direct operated remote air controlled valve
which controls the oil pressure for varying the operating tonnage of
the press. The air to oil operating ratio is such that approximately
75 PSI air pressure will permit the press to deliver 110% rated tonnage.

10) An Air Regulator which provides a remote means of regulating the master
relief valve pressure. The regulator is located at floor level and is
readily accessible to the operator. To change the tonnage of the press,
the operator must vary the regulated air pressure feeding the master
relief valve described in Section 9.

11) An Air Pressure Gage which has two scales on the dial face. One scale
indicates the press tonnage (in tons). The second scale represents
the air pressure being supplied on the pilot line to control the pressure
of the master relief valve described in Section 9.

12) "Ready To Run" and "Slide Overload" Pressure Switch trips at 600 PSI to
indicate the press is ready to run, and resets at 400 PSI to indicate
an overload. When this switch trips, a light on the press master control

panel is illuminated indicating the desired tonnage rating and pressure had been reached in the system. The solenoid controlled valve, described in Section 5, is then de-energized, thereby unloading the bulk of the pump's volume output into the lubrication header. When an overload occurs, the overload relief valve on any corner of the slide drops the supply pressure. When the pressure in the supply line drops below 400 PSI, the pressure switch opens which stops the hydraulic overload pump motor, inactivates the press controls, and immediately sets the brake bringing the press to a halt.

13) Safety "Pop-Off" Valve is in the air pilot line for the master relief valve to limit the maximum pilot pressure available for the hydraulic system. This limit is 110% of rated tonnage. The setting of this valve is determined by design and sealed.

Operation. In the work stroke of the press, the oil pressure under the connection is increased and can exceed the system pressure because the ball-check valve in the overload valve spool prevents the oil trapped under the connection from returning to the system. When the pressure under the connection exceeds the system pressure by 33%, the product of the pressure times the smaller area on the nose of the hydraulic relief valve spool will then equal the product of the system pressure times the larger area on the rear of the hydraulic relief valve spool and the system reaches the point where it is ready to sense an overload. As soon as the oil pressure under the connection exceeds the system pressure by more than 33%, the valve spool will move to the rear because it is overbalanced by the oil pressure under the connection. When the spool moves back, it very rapidly opens large discharge ports so that the full volume of oil under the connection can be very quickly dumped to relieve the press overload. The purpose of using differential areas on the overload relief valve spool is to prevent it from breathing during the work stroke of the press and to maintain hydraulic stability so the system does not prematurely sense, or falsely indicate overload conditions.

The spool is designed so that a rearward movement also will unload the power unit supply source. Because of common connections, simultaneous unloading of all the corners of the slide results even if only one corner is overloaded. At this time the pump supply pressure which feeds the slide, unloads thru the 3/8 P.T. drain port. The pressure drop in the supply line opens the contacts on the "overload pressure switch" referred to in Section 6. This disengages the press clutch and immediately sets the press brake bringing the press to a halt.

With the above operation the pressure gage in the control manifold indicates only system pressure and the preload pressure under the connection. The actual hydraulic working pressure under the connection can exceed the system pressure depending on the tonnage of the work but cannot exceed a maximum of 33% more than the system pressure at any time because of the differential area on the valve spool.

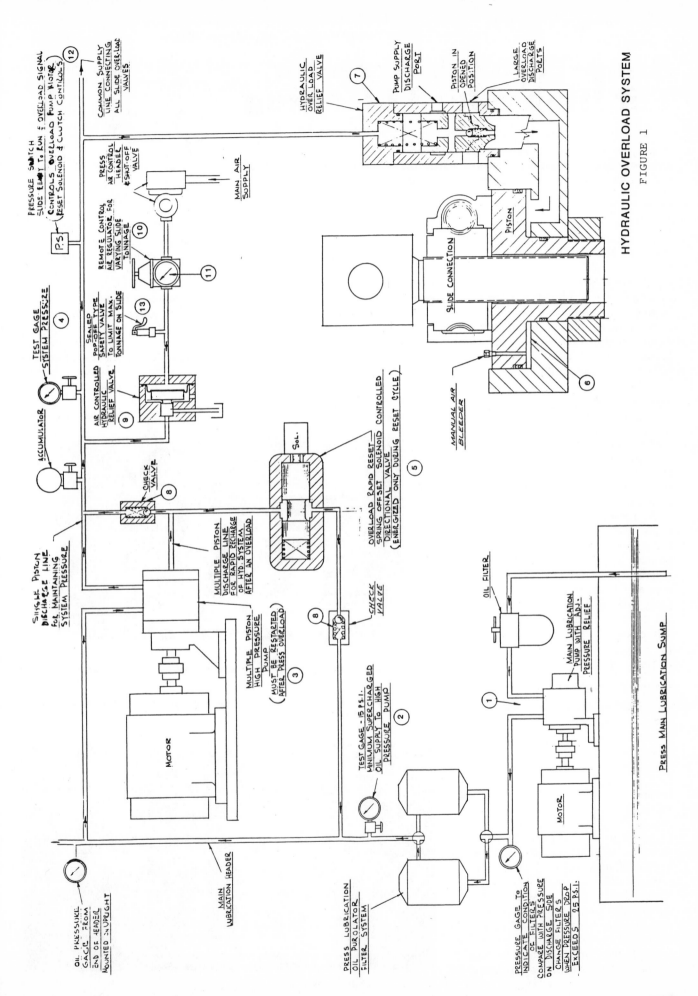

HYDRAULIC OVERLOAD SYSTEM

FIGURE 1

Reprinted from Modern Machine Shop, January, 1980

Snap-Through Arrestor For High-Speed Blanking

A major cause of press breakdown and a major source of noise are associated with the press-loading phenomenon called reversal loads. A system has been developed to reduce these loads.

By THEODORE BOOP, Chief Engineer—Presses
E.W. Bliss Division
Gulf + Western Manufacturing Company
Hastings, Michigan

For many years, manufacturers of high speed blanking presses have been coping with one of the most critical press-loading phenomenon that can exist—reversal loads. Reversal loads, sometimes referred to as snap-through loads, are probably the major cause of press breakdown and a major noise-producing source. Yet, most tool and die people, most metal stampers, and many press manufacturers never calculate and seldom even consider snap-through loads when making their job-to-press application analysis.

With blanking presses being run at higher and higher operating speeds, the snap-through load will continue to take its toll—first through loose bolts and fasteners, then worn bearings and loose linkages, and eventually fatigue failures—unless total press design loading is considered with attention being given specifically to reversal load absorption or dampening.

Cause of Reversal Loads

A brief analysis of the cause of reversal loads will reveal why the phenomenon is so damaging. Visualize the mechanical press as a giant spring which has been loaded to a highly stressed condition by the resistance of the workpiece. During a blanking operation, the workpiece material begins to shear; but before 20 to 30 percent of punch penetration has taken place, the material suddenly breaks through. This breakthrough phenomenon suddenly releases the highly stressed "giant spring" (the press) and the released energy propels the slide downward at speeds of 50 to 100 feet per second. At the extreme end of the slide travel the press members experience reverse loads—connections are in tension instead of compression, and the frame is in compression instead of tension. Then a fluctuation takes place as the press members dampen out the natural shock waves. The harder the material being blanked, the less the punch penetration, and the greater the effect of the breakthrough phenomenon.

A typical blanking press load versus time is shown in Figure 1 with an accompanying snap-through load. An analysis of the press load curve reveals that some kind of force is needed to restrict the "propelled slide" at the point of breakthrough. This restriction would absorb or eliminate the reversed load effect, thus also eliminating the resultant damage to the press. With the reversal load dampened, the desired press-load-versus-time curve would appear as shown in Figure 2.

Research

Experiments were made with springs, rubber pads, air cushions, fixed-orifice-type hydraulics, and nitrogen cylinders with only slight success. These are basically position-sensitive devices which require extremely accurate settings to be 100 percent effective. In addition, they require adjustment when there is a

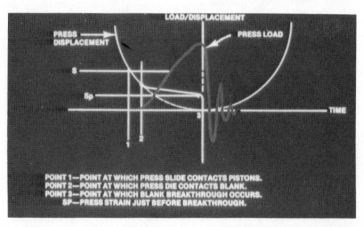

Fig. 1—A typical blanking press load-versus-time curve with no shock dampening.

Fig. 2—If the reversal load were dampened, the press load-versus-time curve would probably have this configuration.

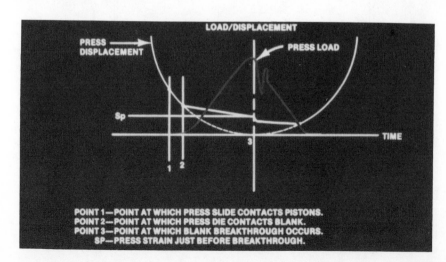

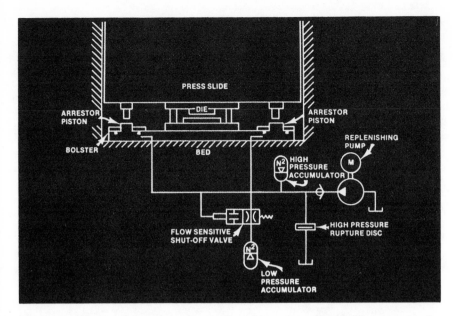

Fig. 3—A snap-through arrestor utilizes high- and low-pressure hydraulic systems to absorb the shock from reversal loads.

change in material thickness. These devices can also exert high loads during the work stroke; therefore, larger press flywheels would be required to supply the additional energy needed to compress them.

Accordingly, the E.W. Bliss Company began the development of a snap-through arrestor to effectively dampen the noise and load effects of press reversal loads. Such a device would eliminate the damage that can result from snap-through loads that occur during a high-speed blanking press application.

The Snap-Through Arrestor

The Bliss snap-through shock arrestor utilizes hydraulic cylinders incorporated in the press bolster. The patent-pending system, shown in Figure 3, immediately senses an increase in fluid velocity when blanking breakthrough is initiated. A valve is actuated in less than 0.001 second, shifting the system from a low pressure mode to a high pressure mode, and retaining all the fluid in the cylinders to provide maximum restraining force. The result is much greater efficiency in reducing shock and noise, and practically no heat generation in the fluid.

The hydraulic cylinder pistons must be engaged by the slide actuator stems just before breakthrough of the blank. The normal slide velocity (possibly 16.457 inches per second) will cause only minimal pressure drop through the flow-sensitive valve and the system functions with the low-pressure accumulator (50 to 100 psi) as shown in Figure 4.

Acceleration of the slide at breakthrough will immediately close the velocity-sensitive valve. A rapid counter-loading is then produced for retarding the energy released from the propelled slide. A high-pressure accumulator cushions the system in this mode and maintains the fluid in the system as shown in Figure 5.

The basic system design parameters require a minimum volume of fluid in the cylinders and pipes, a fluid with high bulk modulus, and a rapid response of the velocity-sensitive valve.

The press load build-up cycle, the load-release cycle with no dampening, and the load-release cycle with snap-through dampening are shown in Figure 6. The snap-through arrestor system is designed for a 70

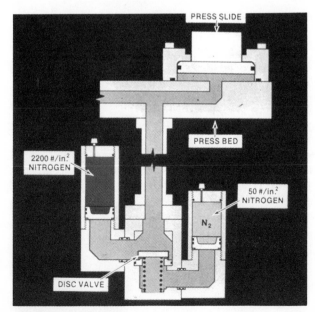

Fig. 4—At the instant of breakthrough, the low-pressure hydraulic system is employed.

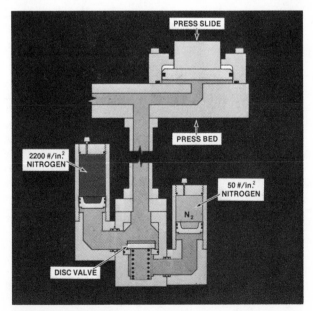

Fig. 5—Immediately after breakthrough, the high-pressure hydraulic system comes into play.

173

percent reduction in the released energy due to breakthrough. A reduction in the snap-through shock loads in high-speed blanking presses will extend the normal life of the press, maintain the closely held clearances, reduce costly maintenance, and increase die life.

Reduced Noise

Blanking operations create noise at three stages:
1. At point of impact between the die punch and the metal to be blanked.
2. At point of metal fracture.
3. At point of bottoming-out of the press connection clearances.

The snap-through arrestor has a sound dampening effect since it attacks the major noise "source" points in press blanking by restricting the effects of the metal fracture and by eliminating the bottoming-out effects. **MMS**

Fig. 6—The snap-through arrestor system is designed to reduce the released energy due to breakthrough by 70 percent.

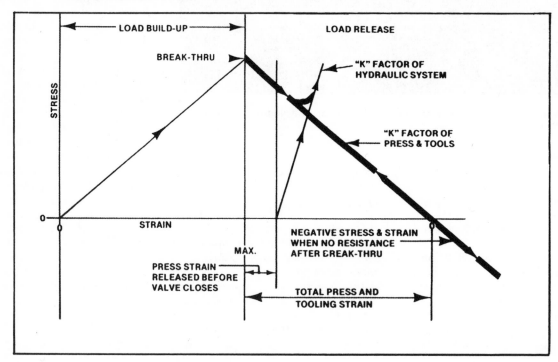

Editor's Note:

The snap-through arrestor received its initial field test on blanking presses at the Denver Mint. It has lived up to its original expectations. Other units are now being installed for further testing. It is expected that the concept will be on display at the International Machine Tool Show in Chicago and will be commercially available, at that time.

The heart of the system is the velocity-sensitive valve. The valve must be designed and manufactured for a specific set of velocity conditions. Since each blanking job involves different presses, materials, thicknesses, and so on, the breakthrough velocities are different from one job to the next. Thus, to go from one job to the next it is necessary to change to the appropriate valve for the job. The valves are not adjustable.

CHAPTER 5

NEW STRIDES IN PRODUCTIVITY

Reprinted from Machine and Tool Blue Book, November, 1979

Press Advances Spur Stamping Productivity Gains

New developments in press technology are providing the metal stamping industry with safer, more reliable, more productive machines. Automated systems boast continuously running presses, automatic parts transfer, quick die change and other features designed to produce more parts per hour with lower labor costs.

By LEO R. RAKOWSKI, Engineering Editor

Higher labor and material costs, need to improve productivity to stay competitive, operator safety and plant noise limitations are just a few of the challenges facing captive and contract metal stamping operations today. Increasingly, metal stampers are looking to manufacturers of presses and auxiliary equipment for help in solving these problems. And the builders are responding with a steady stream of technological improvements which is moving the stamping industry into the age of automation. The cross section of press improvements described in the following pages is representative of the greater accuracy, efficiency, productivity and safety being engineered into today's stamping presses.

Curbing Destructive Forces

Stamping presses literally try to tear themselves apart, remarked one press manufacturer. To improve press and die life and minimize press maintenance, press builders have added numerous refinements to balance press forces and to permit more precise control of moving elements.

For example, Niagara Machine and Tool Works, Buffalo, N.Y., produces the BIF Series of automatic, high-production, progressive die presses. The acronym stands for Balanced Inertia Forces, and represents a press design that not only balances all revolving inertia forces but opposes the reciprocating forces of the slide. The result: greater life for press members and less maintenance traceable to vibration-related problems.

Niagara's BIF Series was recently modified to include an improved Die-Saver Lock. This feature eliminates all clearances in the press' slide adjustment mechanism, minimizing vertical clearances to improve press performance and extend die life through optimum control of punch penetration.

The BIF press also has an automatic circulating oil system which delivers oil to all of the bearings in the crown, gears, slide barrel adjustment, gibs and counterbalance cylinders. Continuous pressure lubrication permits a precision fit of bearings and gibs, precise alignment, longer life for bearing surfaces and longer die life. An electrical interlock on the lubricating system automatically stops the press if the pressure changes or if the oil level drops.

Die Breakthrough Breakthrough

In conventional metal stamping operations, when the punch penetrates to about one-third the thickness of the metal being stamped, the metal shears, suddenly removing all resistance to the downward movement of the press slide. This phenomenon is called die breakthrough, and the sudden acceleration of the slide at the bottom of its stroke creates tremendous loads which lead to loose fasteners, worn bearings and linkages and, eventually, fatigue failures in vital moving parts.

Recently, press manufacturers have developed systems which relieve the downward force on the slide at the moment of die breakthrough to prevent the buildup of potentially harmful loads. For example, a 1000-ton hydraulic blanking press designed and built by Erie Press Systems, Erie, Pa., incorporates a pair of "catching cylinders" in its bed which acts as a shock absorber for the main ram.

At the moment of die breakthrough, hydraulic pressure is diverted from the main ram to the catching cylinders by a fast-acting, pressure-sensitive valve to automatically stop the ram with practically no overstroke. The ram stops in less than 0.1 second, without any bump. The catching cylinders dissipate the energy of 1000 tons of force and

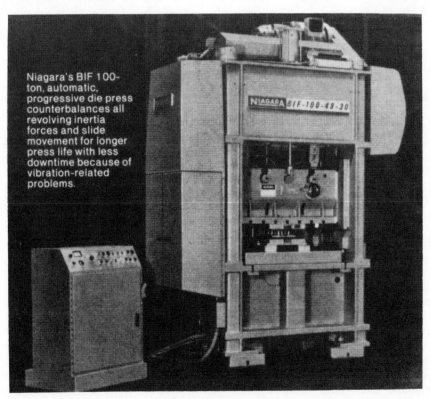

Niagara's BIF 100-ton, automatic, progressive die press counterbalances all revolving inertia forces and slide movement for longer press life with less downtime because of vibration-related problems.

3000 psi oil pressure. To prevent damage to the piping and hydraulic components from shock waves caused by the sudden transfer of pressure, the system includes a shock dissipator.

Erie built the press for a stamping firm which was experiencing mainte-

To get the higher tonnage and greater rigidity required for transfer and progressive die stamping, more and more stampers are opting for straight-side presses, like this 200-ton double crank model made by Rousselle Corp., Chicago, Ill.

nance and wear problems with mechanical presses being used to produce large, washer-like, clutch plate components for automatic transmissions. With the new press, the stamping firm reports a 30 percent reduction in press maintenance costs together with increased die life.

Another system, called a Snap-Thru Arrestor and developed to provide die-breakthrough cushioning for high-speed, mechanical blanking presses, was described by Theodore A. Boop, chief engineer—presses for E.W. Bliss Div., Hastings, Mich., at the Sixth International Congress on Sheet Metal Work held earlier this year in Chicago. Boop termed snap-through, or the reversal loads imposed on the press by the sudden energy release that occurs when the punch breaks through the material, the major cause of press breakdown as well as a major source of noise.

The key to the patented Bliss system is a fast-acting, velocity-sensitive, shut-off valve which cushions the blanking load with a low-pressure mode, but cushions the snap-through mode with a high-pressure mode. Two hydraulic cylinders built into the press bolster are connected to a low pressure accumulator through the shut-off valve, and to a high-pressure accumulator.

The cylinder pistons are contacted by fingers mounted on the slide about the time that the die contacts the material. The normal velocity of the slide as the die penetrates the material causes only a slight pressure drop through the valve so that it remains open and the system functions in the low-pressure mode. At breakthrough, acceleration of the slide immediately (less than 0.001 second) closes the valve, shifting the system to the high-pressure mode to absorb the suddenly released energy of the slide.

The system absorbs about 70 percent of the released energy and promises the following gains for a mechanical blanking press: increased press life, better retention of close press clearances, less maintenance, longer die life, and quieter operation since the noise created at the point of metal fracture and at the point where the slide bottoms out is reduced.

A number of press builders have added eddy current-type variable speed drives which permit precise adjustment of slide speed for maximum production from each press setup. For example, Niagara's PD2 line of gap frame presses, developed for automated, high-production operations using low-tonnage, narrow progressive dies, employs a remote-controlled, eddy-current, adjustable speed drive which permits the operating speed to be adjusted for optimum efficiency according to the die, material feed length and parts ejection, as well as for inching at low speeds during die setting.

The PD2's eddy-current drive, with solid state control, offers a broader speed range, is more compact and requires less maintenance than a mechanical variable speed drive.

Overload Protection

Many presses feature improved systems for preventing press overloads caused by improper press settings, faulty part transfer within the die, double blank and other feed irregularities. The systems prevent the press from exceeding a preset tonnage, stopping the press before it or the tooling can be damaged.

Overload protection can be designed into the press clutch, or incorporated into its hydraulic system. The hydraulic overload protection offered on the straight side presses is made by Danly Machine Corp., Chicago, Ill., an example of the latter type. A hydraulic overload cylinder is built into each of the

Model 27 Speed Stamper high-speed, gap frame press, made by L & J Press Corp., Elkhart, Ind., is designed to turn out small, intricate parts at rates to 800 spm. It boasts such features as thermal sensors to monitor bearing temperatures, eddy current variable speed drive, and cam roll feed capable of handling strip up to 3 inches wide.

press' slide connections. Each has a separate overload and relief valve. When a corner of the press becomes overloaded, the overload cylinder valve for that connection automatically reduces the oil supply pressure, unloading the connection and returning oil to the supply tank. When the pressure in the oil supply line drops below a certain level, the slide overload pressure switch opens, inactivating the clutch and braking the press to a halt. All of the slide connections are interdependent; overloading on a single corner of the slide simultaneously causes the valves on the other connections to unload as well.

Solid State Clutch Control

Availability of solid state clutch controls in the mechanical presses built by U.S. Industries, Inc.'s Clearing division, Chicago, Ill., is still another relatively new press development. Offered as an option, it reportedly matches or exceeds the performance, capability, service life and flexibility of the standard electromechanical control.

The solid state clutch control is more compact than its electromechanical counterpart, and its circuitry can be combined with a solid state automation control to provide a complete logic system for auxiliary press feed cycles, lift cylinders, load monitors, accoustical die monitor and other press operations.

The solid state control can incorporate diagnostic, self-check circuits more easily and at less cost than electromechanical systems. It has the advantage inherent in all solid state controls of no moving parts, a feature which virtually precludes maintenance problems. Once the system is operational, long service life can be expected.

While the press builders continue to improve the efficiency and reliability of their products, metal stampers are eagerly utilizing a host of productivity-boosting features. To minimize labor content, they are consolidating part forming operations in a single, continuously operating press, equipped with progressive or transfer dies.

There are a number of prerequisites to the use of such dies: first, the bed of the press must be large enough to accommodate the tooling; the press must be accurate and rigid enough, and have operating safeguards adequate to protect the considerable tooling investment; the press must be capable of continuous operation at speeds sufficient to utilize the productive capability of the tooling; finally progressive die parts (and increasingly transfer die parts) are produced from strip, requiring the addition of an uncoiler, stock straightener, precision feed and lubricator and some provision for handling of completed parts.

The advantages gained with progressive and transfer die stamping justify the investment, however. In the case of labor savings, only a single press operator is required since the part is completely formed in the press. Since a completely finished part is produced with each stroke of the press, secondary operations and multiple handlings are eliminated.

Press utilization is high. Automatic coil feed systems permit continuous operation with downtime only for threading new coils and die maintenance. The operator is no longer required to load, operate and unload the press, minimizing the opportunity for mistakes. At the same time, the die area can be completely guarded, eliminating the danger of hands-in-the-die-type accidents. Since the operator is

effectively removed from the immediate die area, the dies can be totally enclosed which presents a greater opportunity for muffling press noise.

More Straight-Side Presses

Increasing sales of heavier-tonnage presses with larger beds suggest that metal stampers are increasingly exploiting the productivity-boosting potential of progressive and transfer die stamping. Maurie Gallaher, vice president—marketing for South Bend-Johnson presses made by South Bend Lathe, Inc., South Bend, Ind., reports that, "We sell more OBI presses than any other type because of their lower price and shorter delivery time. However, our straight-side press sales have been growing at a faster rate.

"The trend is to concentrate multiple-operation stamping jobs on a single press using progressive or transfer dies. Straight-side presses, because of their greater rigidity, superior alignment, reduced deflection and greater tonnage capacity, provide the best results for these stamping operations. We also see increasing use of automatic feed and loading and unloading equipment," Gallaher reported.

Adam Schwarz, vice president, engineering for Rousselle Corp., Chi-

South Bend-Johnson 200-ton straight-side press features automatic roll feed, stock oiler and scrap chopper options.

cago, Ill., confirms the trend to larger presses: "Job shops, which were formerly characterized by single operations on small presses, have combined these operations—where part volume permits—on single, larger presses.

"Instead of using three or four

small presses, three or four die sets and feeds and employing three or four operators, job shops are producing parts requiring multiple operations in a single work center which increases productivity, and reduces handling and precious plant space requirements.

"Small presses will continue to be in demand and will account for most of our sales volume, but introduction of our larger presses has opened the way for new applications, resulting in increased sales. Where five years ago we produced relatively few presses over 150 tons, today our sales of these presses have increased considerably," Schwarz revealed.

Since raw material costs account for about half of the total cost of the part, metal stampers are also concerned with getting the maximum number of parts per sheet or coil. Feeds recently developed to improve material utilization include the Stagger Feed system offered by Waterbury Farrel Div. of Textron Inc., Cheshire, Conn., manufacturer of transfer presses.

The unit, a mechanical feed which quickly attaches to any transfer press and which can be moved from one machine to another, can be used for single-row blanking or can automatically shift the strip from side to side for double-row blanking. In the latter mode, two rows are blanked in a single pass through the press, eliminating the need for a takeup reel and reducing scrap loss by 7 percent.

Die-Changing Time Slashed

Perhaps the largest single contributor to lost production time on stamping presses is the amount of time required to change dies. Installing and removing dies can exceed the running time required to produce short-run quantities of a part, more than doubling the machine hours charged to the job.

One solution to this problem is the Quick Die Change (QDC) method developed by Danly Machine Corp. for large, heavy tonnage, straight-side presses of the type normally found in automotive and appliance plants (QDC is a registered trademark of Danly Machine Corp.).

The QDC system replaces the conventional press bed with a pair of track-mounted, self-propelled die carriers, each containing a complete die. One die carrier is positioned in the press; the other is conveniently positioned alongside the press, permitting die-setting for the next run without interfering with production.

At the conclusion of a part run, the die carrier in the press is disengaged and moves out of the press through an opening in the press upright. The other die carrier moves into the press through the upright on the opposite side. After the die carrier is positioned in alignment with the press slide, its wheels retract. Die clamps mounted on the press slide engage slots in the upper half of the carrier to secure the upper die to the slide, and the press is ready for production.

The press user can opt for front-to-back or side-to-side movement of the carriers through the press as part requirements or space limitations dictate. Or the die carriers can be replaced by a rolling bolster and a crane.

Biggest advantage of the QDC system is the time savings. Downtime for tool change on the large press is reduced from hours to a few minutes. Die-setting is completely handled off the press, freeing still more press time for production.

With QDC, short runs become more economical, try-out of dies on the production presses is more practical, and production schedules can be more closely tailored to assembly-line needs for more efficient inventory levels.

Rapid die change is not exclusive to large presses. Minster Machine Co., Minster, Ohio, has offered the feature on its Die-Namic OBI and straight-side presses for years and, more recently, has incorporated the feature in its Die-Namic TranSystem straight-side presses for transfer stamping operations.

In the Minster system, the dies are mounted on standard upper and lower die plates. The die plates mount in pneumatic fixtures in the press which automatically clamp the dies into alignment. Shut height for the press is marked on the die; the operator simply resets the shut height by means of the power slide adjustment until the proper reading registers on the press shut-height indicator. With Die-Namic, dies can be changed in one-sixth the time of a conventional die.

The standard die plate principle is used in Minster's TranSystem. The press has a four-post master set with fixtures for the individual die sets. The press can be provided with three to eight tooling stations depending on the customer's requirements.

Tooling is simple compared to trans-

fer dies designed for use in standard straight-side presses. Blank or strip feed and transfer finger rails are independent of the tooling, reducing costs to about half that for a conventional transfer die. The modularity of the tooling permits one or two stations to be changed easily and rapidly, enhancing TranSystem's use for family-of-parts production.

With TranSystem, the press can be completely retooled—including stock

Danly's Synchro-Matic press line, with automatic loading, unloading and conveying, permits automated stamping of panels at rates to 1200 per hour. Quick Die Change press feature reduces tool changeover time for the huge presses to minutes, facilitating regular rotation of part production.

feed and transfer finger bars—in less than one hour. The relatively short setup time enables the press user to schedule runs for quantities as small as 1000 pieces.

Previously, tooling and setup costs precluded the use of the transfer process for small-volume runs of parts requiring multiple operations. TranSystem not only makes the transfer process more practical for short runs, but eliminates the danger of hand-feeding parts through multiple press operations.

A Systems Approach

While numerous improvements in stamping presses have been made, exploiting these improvements, and those still to come, dictates the use of compatible auxiliary equipment. Stock

feeds must match press speeds, loading and unloading devices must be reliable yet flexible enough to permit rapid job changes . . . in short, the system is no more productive than its most inefficient component. This "systems approach" is common to the following examples of the current state-of-the-art in metal stamping.

Danly Machine Corp.'s Synchro-Matic press system is a fully automated, synchronized line of several presses that takes in blanks at one end and turns out finished parts at the other—without the aid of operators. Each press has an automatic loader and unloader which move the part through the press, and a conveyor which automatically shuttles the part to the next press in line.

Loader, unloader and conveyor are mechanically driven by the press they serve. The loader and unloader have arms which terminate in one or more grippers shaped to match the part being run. Movement of the arms and the opening and closing of the grippers are controlled by cams located in the loader and unloader housings.

The arms have a quick-disconnect feature designed for fast replacement of arms with grippers matched to the next job to be run. Similarly, each conveyor has a nest tailored to the shape of the part being run. Conveyor side rails are adjustable for part width, and loaders, unloaders and conveyors can be adjusted to suit different die heights.

"The need to reduce labor content is the greatest single problem facing the metal stamping industry," notes John R. Danly, vice president, sales, for Danly. "In the auto industry, for example, for years the normal press line consisted of manually-loaded presses fitted with mechanical extractors and between-press conveyors. More recent installations include automatic de-stackers which load the lead press. Dies for the subsequent-operation presses have guides, gages and stops which permit hands-out-of-the-die loading and achieve higher line speeds.

"However, these lines still require an operator for each press and are limited in output because each press runs on a single-stroke basis. A fully automatic press line, like the Synchro-Matic, significantly reduces labor content since individual operators are not required for the presses—the line runs virtually by itself.

"Continuous operation of the

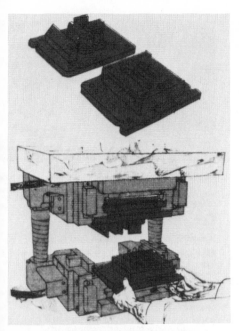

Minster's Die-Namic system reduces die change time to 90 seconds for an OBI and to under an hour for a transfer press. Dies are mounted on standard upper and lower die plates (top) which insert into pneumatic fixtures in press. Dies are automatically secured in precise alignment.

presses, another important feature, increases productivity by one-third to one-half. Our automatic press lines are turning out automotive panels at production rates up to 1200 panels per hour. I don't know of a manually fed line that achieves half that rate, yet the press speeds for both types of lines are about the same. The greater productivity of the automated line derives from the continuous running of the presses. Another benefit of the continuous operation is the elimination of wear problems associated with constant starting and stopping.

"Still another important feature of the automated press line is its flexibility. More recent installations have readily changeable loader and unloader stroke length and motion, readily changeable panel strippers and panel nests on conveyors, etc. Lines of this type are producing up to 12 similarly-sized stampings on a regular rotating basis," Danly notes.

An example of automated stamping at the other end of the part size spectrum is the Hummingbird, a progressive die press developed by Minster Machine for high-speed blanking and shallow-forming. Available in 30 to 60-ton capacities in high-speed and ultra-high-speed models, the 30-ton UHS press is capable of speeds to 2000 spm

with a stroke of just under one inch.

Keys to the press' exceptional stroking rate are a patented mechanical drive, pressurized lubrication system, a hydrostatic bearing system which eliminates lateral slide movement, and an hydraulic bed lock which eliminates all clearance in shut-height adjustment. The press is 100 percent dynamically balanced so that no foundation is required.

Fully utilizing the Hummingbird's productive capability dictates the use of performance-matched ancillary equipment. Minster offers the press separately, or as part of a complete package which includes as optional equipment: a high-speed, cam-operated roller feed accurate to 0.001 inch per inch of feed—and designed to stop instantly in the event of die fault or a stock buckle; a loop control system which supplies coil to the press feed at 4800 inches per minute; a straightener; and a double-coil reel.

One customer, who is using the Hummingbird to blank and punch laminations for electrical components at 1600 spm, reported that the high-speed press provides five times the productive capability of the two automatic straight side presses it replaced. The customer expects to get three million hits between grinds from the carbide progressive dies, or twice as many hits between grinds as achieved with the straight side presses. Dimensional tolerances for the laminations produced are typically 0.002 inch, but 0.0005-inch tolerances can reportedly be held because of the slide's precision guidance system.

The press' die area is completely enclosed, eliminating the hazard of hands-in-the-die accidents for the operator. And the press is so compact that it can be completely enclosed to reduce noise.

"Over the past 20 years, our customers have resorted increasingly to automatic stamping presses—both progressive and transfer die presses—to increase productivity. Automation has not only permitted the higher press speeds needed but has also been the key to increased operator safety," explained Merrill L. Ridgway, vice president, marketing for Minster.

"As operating speeds have increased, customers have had to be more concerned with other components of the system. Handling the material, feeding it into the press, and re-

Platarg Engineering's Model 611 cam-operated, multiple square plunger transfer press has 12 forming stations and can automatically produce drawn or intricately formed parts at rates to 90 spm.

moving finished parts and scrap from the press are all important considerations.

"Last but not least is the need for tooling that will hold up under the higher speeds. The greater productive capabilities of high-speed presses cannot be achieved if the machines must be stopped every 10 hours or so for tool maintenance. In fact, if the tooling cannot hold up, the ratio of downtime for tool maintenance and changeover to the next job increases with higher press speeds.

"That is why quick die change systems, like our Die-Namic system, are important. They not only fractionalize the time required to change dies, but in the process improve the economics of short-run production for metal stampers. At last count, over 700 companies are using the Die-Namic system and the market is growing rapidly."

"Faster operating speeds continue to be important," Ridgway continued. "Progressive die presses are being used to produce larger parts. Today, 300 to 400-ton presses are running at 300 to 400 spm—which is 50 to 100 percent faster than previously obtained.

"At the same time, parts are being produced from heavier gage materials. For example, motor mounts for the auto industry are being produced on progressive die presses at rates up to 100 spm," Ridgway notes.

The C (Cam) and E (Eccentric) Series transfer presses built by Platarg Engineering Ltd., London, and distributed in this country by Platarg Engineering Inc., Branford, Conn., are another example of multiple-station machines designed to produce completely finished parts from coil.

Used for shell-type parts and parts which require multiple piercing and drawing operations, the C Series presses have from 10 to 13 stations and will produce parts drawn to a depth of over 2 inches from blanks just under 4 inches in diameter at speeds up to 160 spm. The E Series, for larger parts, offers 5 to 11 stations, depending on the model, and will produce drawn parts from 6-inch-square blanks at rates up to 105 spm.

On both press series, each forming

Minster's Hummingbird progressive die press incorporates a number of advanced design features, enabling part production to close tolerances at speeds to 2000 spm.

station has its own cam-driven ram. Each stage is individually adjustable for stroke and shut height. The transfer mechanism is built into the press, so that tooling for the individual stations is no more complicated than that used for single-operation presses. Tool stations can be removed individually to accommodate part design modifications, for family-of-parts production and for easy tool maintenance.

Never Lets Go

A problem area for transfer presses in the past has been positive transfer of parts at high operating speeds. Platarg has solved this problem with a system that maintains constant control of the part through the press. The part is

never unsupported. As an added precaution, a solid state monitoring system automatically disengages the clutch in the event of a part transfer problem, stopping the press before the tooling can be damaged.

The Platarg transfer presses offer a number of important advantages over part production on a battery of single-operation presses, for example: Production is achieved with a single, compact machine. A finished part drops off with every stroke of the press. The press readily accepts accessory equipment for thread rolling, beading, side piercing, etc., eliminating the need for secondary operations.

Part production is continuous and automatic, eliminating intermediate handling, inspection and storage. And since forming operations can be spread over many stations, intermediate annealing of drawn parts is also eliminated.

Coil feed and automatic transfer of the part through the press make for maximum operator safety. Tooling is completely enclosed and entry into the die area triggers a safety device which stops the press.

Material utilization rate is high since strip can be slit to precise part width requirements. Extra strip width required for pilot holes or for a skeleton to carry the part through the forming operations as in progressive stamping, is not required.

Previously, the complexity and high cost of tooling for transfer stamping has limited application of the process to high-volume production. The simplicity of the tooling and the ease with which it can be replaced have reduced tool costs and press downtime for job changeovers to the point where the process is becoming more economical for smaller runs.

Turnkey System

One of the most recent developments in press technology is the supply of complete turnkey metal stamping systems by U.S. Industries, Inc.'s Clearing division, Chicago, Ill. One system, dubbed Panomax and built for one of the big-three automakers, will produce quarter panels, fenders and similar body panels at a rate of 600 parts per hour, as against 450 panels per hour for a more conventional press line.

A departure from a conventional

Flexopress, made by Teledyne Precision-Cincinnati, Cincinnati, Ohio is shown making easy-open lids for beverage cans. Press boasts a gibless precision slide which runs on ball-bearing raceways and such control features as a press-mounted load limit detector and a bearing temperature monitoring system.

press line of six or seven presses, Panomax consists of a 1000-ton, double-action press in tandem with a 1600-ton, two-slide, tri-axis transfer press with a 72-inch feed stroke thought to be the largest in existence.

Movement of parts through the two presses is completely automatic: a destacker feeds sheet blanks one at a time into the double-action press; the drawn blank is automatically transferred to a turnover mechanism which loads the blank into the transfer press; the part is conveyed through five work stations, two on the first slide and three on the second; from the final forming station the finished parts are loaded on a conveyor and moved from the press area. Along with the presses and conveyors, Clearing furnishes the material handling equipment, sheet loaders, coil feeds, unloaders, stackers and destackers and other components that comprise the system.

Panomax represents a number of important developments. First, turnkey responsibility allows the press builder to innovate since the constraints imposed by compatibility with ancillary equipment the customer might choose to add are removed. Second, the press builder can design the system for maximum output, selecting components for optimum effectiveness.

Panomax achieves a higher production rate with two presses than a conventional line of six or seven presses.

Panomax, a complete stamping system built by U.S. Industries' Clearing division for a major automaker, automatically produces 600 auto panels per hour. The system, which consists of a 100-ton double action press in tandem with a 1600-ton, two-slide transfer press, combines the high productivity and compactness of a transfer press with the part size flexibility of a conventional multiple press line.

Labor costs are reduced at the same time that valuable plant space is saved for other operations.

The line offers a degree of flexibility unusual to transfer presses. While the Panomax' double-slide transfer press is tooled for five stations, each slide could be changed to accommodate up to four stations, providing a range of from two to eight forming stations. This provides the Panomax with a part size range extending from small panels and components to large, one-piece body side panels. ● ● ●

[Editor's note: Application of new technology to metalworking presses is perhaps most dramatically demonstrated in the area of punching and notching machines, or turret punch presses. Developments in this area, such as press computer controls, laser hole making capabilities, and modular tooling were covered in detail in BLUE BOOK's July 1978 issue.]

Prepared for *Understanding Presses And Press Operations*

Hydraulic Presses In The 80's

by

Robert G. Lown
Vice President - General Manager
Greenerd Press & Machine Co. Inc.
41 Crown Street
Nashua, New Hampshire   03061
USA

Abstract

This paper summarizes trends in the recent buying patterns
and applications of hydraulic presses in the United States.
It outlines some of the reasons for these changes. Characteristics of hydraulic presses are reviewed. Some guide lines
are offered to assist in making a press selection. Predictions
about changes in the use and construction of hydraulic presses
are presented.

Introduction

The hydraulic press is one of the oldest of the basic machine
tools. As a machine it is probably as old as the bicycle.
And like the bicycle, it is enjoying renewed popularity in the
1980's. Lines of hydraulic presses are showing up in plant
after plant. Electric motor manufacturers assemble motor shafts
to rotors, compress laminations, and press cores into motor housings.
Automotive manufacturers stamp out fuel pump diaphragms, assemble
shock absorbers, and form disc brakes. Aircraft companies form
tough titanium housings and trim forged turbine blades. Firearm
makers broach revolver parts. Jewelers coin Boy Scout pins.
Tuba bells and cymbals are shaped for musical instruments. The
computer industry trims p.c. boards, forms chassis, and folds up
computer bases. These and hundreds of other jobs are being done
today in industry by hydraulic presses.

Why the Trend Towards Hydraulic

The American Machinist, a McGraw-Hill publication, has been gathering machine tool population figures for many years. These figures are widely respected and generally considered as the most authoritative representation of American machine tool population, age, and industry distribution. These reports show that up until a few years ago, less than one out of 10 presses in American industry were hydraulic presses. According to the latest survey this figure has risen to about 1 out of every 4. The more recent Federal government figures of shipments of new hydraulic presses in the last few years confirm this trend.

Why should a product that has been "around" as long as the hydraulic press suddenly seem to take on a more important role. There are a variety of reasons. First of all it should be recognized that other types of presses have improved their speed and overall productivity. Fewer presses, regardless of the type, are needed to accomplish the same or even more work. This fact tends to distort the figures.

Recognition of the capabilities of the hydraulic press has come along slowly. The mechanical press has dominated the thinking of press users for at least 60 or 70 years. The term "punch press" has almost become synonomous with the term mechanical press. The training of tool and die makers and manufacturing engineers has almost always been oriented towards mechanical presses. The press room terms used, the approach to tooling, feeding, and part handling have all been centered around the mechanical press. All of this has been true while the hydraulic press of old was properly regarded as slow...and leaky.

But hydraulic presses have made substantial improvements in performance and reliability. The presence of other types of hydraulic power in the typical industrial plant has reduced some of the fears about the unfamiliar. Maintenance departments know how to service hydraulic components. New fast acting valves, fast acting electrical components, and much faster hydraulic circuits have put some "zip" into the hydraulic presses and moved them out of the maintenance departments right out on to the production lines. These developments have spurred interest in hydraulics.

Cost, too, has been a factor. OSHA has contributed to this aspect. The cost to "OSHA-ize" older presses has become substantial. This has caused many buyers to weigh the relative costs of rebuilding older presses against the purchase of a new machine. In considering new presses more and more buyers include a hydraulic press in their considerations as an alternative to the replacement of a mechanical press. In some cases, if the application is appropriate, there can be significant cost savings with the hydraulic press.

Still another factor has affected this trend towards hydraulics. The marketing of manufactured products in this country has changed. More businesses offer a wider selection of models, sizes, specials, and customized products. The net effect is often shorter runs. These short runs and more frequent setups tend to favor the use of hydraulic presses.

The vast majority of press jobs (perhaps not parts) in this country are still hand fed. Lot sizes tend to be small. These short runs are too small to justify costly automatic feeding systems. The life of the run is too uncertain for many to invest in the "ideal" set-up. So many press users stay with the simpler hand feeding methods. Hydraulic presses can be competitive on most hand fed work and even on automatically fed jobs.

What is a hydraulic press? First and foremost it is a press. As a tool, it does essentially the same kind of work that a mechanical, pneumatic, or drop hammer does. It delivers a controlled force to accomplish work. The method of delivering this force is through the use of hydraulic power. Fluid is controlled, contained and intensified in one or more rams in various directions.

Figure 1 shows a typical hydraulic press. In this illustration the press is in the "C" frame configuration. In this example the electric motor rotates a pump to draw oil out of the oil reservoir. A valving and electrical system respond to commands to advance and retract the slide or ram and apply the amount of desired pressure. The frame acts as a rigid base for resisting movement and a surface to mount tooling.

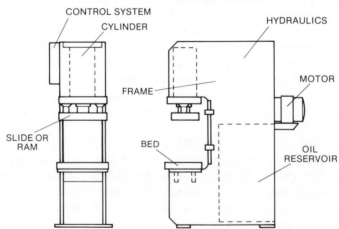

TYPICAL HYDRAULIC PRESS

The construction of a hydraulic press and its characteristics will vary depending on the intended use. Hydraulic presses are built for rubber molding, forging, laminating, compacting, various type of plastic molding, and metalworking operations such as blanking, coining, drawing, swageing, and assembling. The style of the frame and the hydraulic circuit will vary considerably with the use.

In this discussion the emphasis will be on metalworking hydraulic presses since this application is so widespread and the design of a metalworking hydraulic press apply to most other types of presses.

Hydraulic presses have a number of characteristics in common.

Characteristics of Hydraulic Presses

1. Full Power Stroke (Figure 2) The full power of a hydraulic press can be delivered at any point in the stroke. It is not necessary to allow for a reduced tonnage at some point near the bottom of the stroke. This feature allows the user to adjust the stroke only for the purposes of part clearance. A rough setting is accurate enough. When changing die sets of different heights there is no necessity to make a fine adjustment of the ram stroke. The desired (pre-set) pressure will be applied to the work when the tools encounter the resistance of the work. Different tool heights or varying thicknesses of material have no effect on the proper application of pressure.

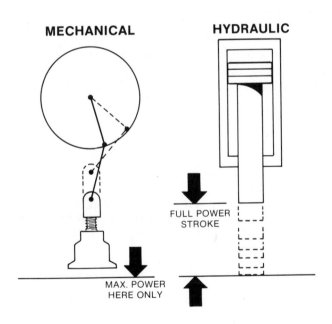

MECHANICAL   HYDRAULIC

FULL POWER
STROKE

MAX. POWER
HERE ONLY

The full power stroke feature is most easily appreciated in the deep drawing application. In those cases the maximum power requirements generally occur when the punch contacts

the work. It continues, at a reduced pressure throughout
the stroke until another power requirement may occur at the
very bottom of the stroke to iron in a radius.

2. Built-in Overload Protection (Figure 3) A 100 ton press,
    for example, will exert only 100 tons of pressure...no more.
    This overload protection is always present throughout the
    stroke. It reacts instantly to an overload and need not be
    reset if such an overload occurs. The overload feature is
    generally provided by one or more hydraulic relief valves
    and backed up by other redundant features.

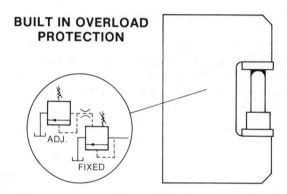

**BUILT IN OVERLOAD
PROTECTION**

3. Low Operating Costs (Figure 4) Hydraulic presses have very
    few moving parts. Those parts that do move operate in a flood
    of pressurized oil...a kind of built-in lubrication system.
    Breakdowns, when they do occur are usually caused by the failure
    of a minor, easily replaceable part such as a ram packing, a
    coil in a solenoid valve, an electrical relay.

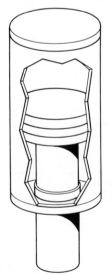

**MAJOR MACHINE COMPONENTS
OPERATE IN A FLOW OF
PRESSURIZED OIL**

Since set-up time tends to be somewhat easier and down time tends to be shorter the production time on hydraulics have a good record with most press users.

4. <u>Large Capacity</u> (Figure 5) The bed size, stroke length, speed, and tonnage of a hydraulic press are not necessarily interdependent. Very large bed sizes of 72" x 48", for example, are available with only 20 tons of pressure. Conversely a 200 ton press can be purchased with a small 36" x 36" bed area. Stroke lengths of 12", 18", 24", and longer are readily available. Since the stroke length is fully adjustable, the full 18" can be used to open the press up wide for the installation of tools. In production, if clearances allow it, the stroke length can be set as close as 1/4". So the elements of capacity vary without regard to each other and most of these elements are adjustable.

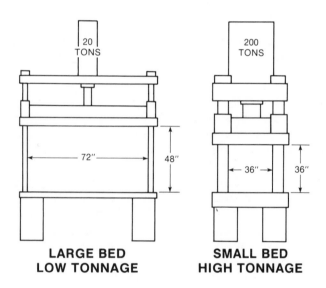

**LARGE BED
LOW TONNAGE**

**SMALL BED
HIGH TONNAGE**

5. <u>Quiet.</u> The nature of the hydraulic press, with a properly designed hydraulic system, offers a surprisingly quiet machine in view of the tasks performed. Decibel readings taken on typical presses find maximum readings in the 85 db range. Hydraulic ram movements can be controlled to pass through the work without impact and with a minimum of shock.

6. <u>Adjustable Tonnage</u> (Figure 6) The tonnage of a hydraulic press can be turned up or turned down. It can be programmed just as the movements of the press can be programmed. The relief valve system that functions as the built-in overload protection serves the dual function of pressure adjustment. This allows a 75 ton press to be set at any point below 75

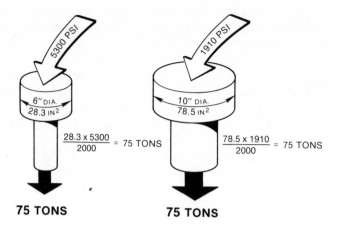

$$\frac{28.3 \times 5300}{2000} = 75 \text{ TONS} \qquad \frac{78.5 \times 1910}{2000} = 75 \text{ TONS}$$

75 TONS **75 TONS**

tons for a particular job. Usually there is a lower limit...
generally about 20%.

Tonnage of a press can be confusing. Figure 6 shows how two
different size pistons both deliver 75 tons of pressure.
The 75 tons is achieved by applying different hydraulic
pressure to the different areas of the piston.

But there is no difference in the force applied to the work.
Each ram can deliver up to 75 tons or 150,000 lbs. of force.
This 150,000 lbs. of force is distributed over the work piece
put into the press. If the work piece happens to present
100 sq./in. of surface, this 75 ton press at full pressure
will exert 1500 psi on that work. If less pressure is desired,
the pressure can be turned down.

7. Hydraulic Press Speeds

Most press users are accustomed to describing press speeds in
terms of strokes per minute. This is easily determined with a
mechanical press and is always part of the specifications of the
press. It is not so easily determined with hydraulic presses.

The number of strokes per minute for a hydraulic press are
determined by calculating a separate time for each phase of the
ram stroke. The rapid advance time is calculated, then the
pressing time, (the work stroke); then, if there is no dwell
time, the rapid return time.

The basic formula for determining the length of time in seconds
for each phase of the stroke:

$$\frac{D}{IPM} \quad x \quad 60 = T$$

D = distance of a particular phase of the stroke in inches
IPM= ram speed of a press in inches per minute
T = time in seconds

Example: a hydraulic press with a 800 IPM rapid advance, 80 IPM pressing speed, and an 800 IPM rapid return. The work requires a 2" advance, a .250" work stroke, and a 2.250" rapid return.

```
Rapid Close    2"   ÷ 800 IPM x 60 = .15 seconds
Press        .250"  ÷  80 IPM x 60 = .19 seconds
Rapid Open 2.250"   ÷ 800 IPM x 60 = .17 seconds
Allowance for valve shift, electrical
reaction, & decompression, etc.     .75* seconds

                 Total         1.26

       60 ÷ 1.26 = 48   cycles per minute
```

*This figure will vary depending on the nature of the work and the hydraulic system. It is a reasonable average for most types of hydraulic presses.

Although this calculation appears to be cumbersome it is necessary to give an accurate answer to the question of speed on a given job.

Hydraulic presses are not yet considered to be in the "high speed" press range. In the automatic mode, however, hydraulic presses are operating in the 20 to 100 stroke per minute range. At least one domestic and certain European machines have achieved higher speeds.

This speed range satisfies the requirements of hand fed work. The speeds are comparable to conventional geared and geared OBI's.

Guide Lines for the Selection of a Hydraulic Press

1. Tonnage: Is the tonnage required to do a job the same for a hydraulic press as it is for a mechanical press? The answer is yes. There is no real difference. The same formulae are used to determine tonnage. There may be certain applications such as deep drawing where the full power stroke characteristic of hydraulic presses tends to reduce the tonnage but there are no known instances where using a hydraulic press required more tonnage.

    The most misleading practice is to assume that because a job is now successfully done on a 100 ton mechanical press that the pressure required is 100 tons. It has to be asked, was it ever tried at 80 tons? At 50 tons? With a hydraulic the precise tonnage required to do a job can be determined by observing the press gauge and the successful completion of a part. That is the proper tonnage at which the press can be set.

2. The Action of the Machine. Even though the tonnage question might be settled, the question of the effect of the stroke on the work is often asked. Is it the same as with a mechanical press?

The answer, again, is yes in most cases. There are some specific limitations. Drop hammers and some mechanical presses seem to do a better job on soft jewelry pieces and impact jobs. The coining action seems sharper if the impact is there. In deep drawing the full power stroke may produce different results (usually better). Otherwise there are very few examples where the application of 100 tons of hydraulic force produces any significant difference in the character of the part given the same tooling.

Shear in the dies will reduce blanking tonnage in the same way it does for mechanicals.

3. The Type of Press Selection.

Open gap presses provide easy access from three sides. 4-column presses insure even pressure distribution. Straight side presses offer the rigidity required for off-center loading in progressive die applications.

The more critical the work and the more demanding the tolerances the purchase of reserve capacity is recommended. That is, if the work requires 42 tons to do the job a 75 ton or even a 100 ton press should be considered.

4. Accessories. Most hydraulic press builders offer a wide array of accessories. These commonly include:

. Distance reversal limit switch...sets the depth of ram stroke for automatic return to the up position.

. Pressure reversal switch...sets the pressure at which the ram returns automatically to the up position.

. Automatic (continuous) cycling...a system for initiating (and stopping) continuous cycling of the press. Used when automatic feeding equipment is employed.

. Dwell timers...hydraulic presses can be made to dwell at pressure on the work. This can reduce the total pressure required for a job. An adjustable timer opens the press after a pre-set dwell period.

. Sliding bolsters and rotary index tables...hydraulic
presses can be built with work positioning devices such
as these. The hydraulic systems often power such accessories and include auxiliary functions such as part ejection.

. Die cushions...hydraulic presses are available with hydraulic
die cushions operated from the main power unit. They have
the advantage of taking up less space than air cushions and
tend to offer smoother, more positive resistance.

. Ejection cylinders or knockouts...these accessories can be
triggered to function on the basis of some pre-determined
position, time, or pressure.

5. Quality. Since the applications for hydraulic presses ranges
all the way from simple hand pumped maintenance presses to
sophisticated multi-thousand ton giants, the range of quality
varies accordingly.

Here are just a few construction points that will provide a
basis for comparison of one machine with another.

a. Frame. Compare the weight if possible. Try to determine
the character of the frame construction...if a weldment,
look at the plate thicknesses, extent of ribbing, and
stress relieving.

b. Cylinder and Slide Construction. How big is the cylinder?
How is it constructed? Who makes it? How serviceable is
it? How is the ram travel guided?

c. Maximum System Pressure. At what psi does the press
deliver full tonnage? The most common range for industrial
presses today is from 1000 psi to 3000 psi. The European
machines tend to be well above these levels. But for the
valves and pumps most readily available in this country the
1000 to 3000 psi is more prevelant.

d. Horsepower. The duration, length, and speed of the pressing
stroke is the major determinant of horsepower. Compare
horsepowers.

e. Speed. Take the trouble to calculate the speed based on
the operations you intend to perform. There can be
significant differences in hydraulic press speeds.

Hydraulic Press Limitations

1. <u>Speed</u>: There are no hydraulic presses today that are as fast as the fastest mechanical presses. If speed is the sole requirement and the feed stroke is relatively short, the mechanical press is usually faster.

2. <u>Stroke Depth</u>. Only a few hydraulic presses are available with an accurate built-in method of limiting the down stroke. Generally this has to be provided in the tooling. If a limit switch is used to determine the bottom...no matter if the digital readout stroke setting indicates an accuracy within .001"...the stroke depth is not likely to be controlled much closer than .020". Hydraulic presses are often arranged to reverse at a pre-set pressure. This usually results in very uniform parts. Where absolute accuracy is required "kiss" blocks should be provided in the dies.

3. <u>Automatic Feeding Equipment</u>. The hydraulic press requires some external or auxiliary power to feed stock. The feeder must have its own power and integration of the feeder with the press control system must be accomplished.

   There are, however, an increasing selection of self-powered feeding systems available...roll feeds, hitch feeds, and air feeds.

4. <u>Shock after Breakthrough in Blanking</u>. Both mechanical and hydraulic presses have the same problem. But in addition to mechanical shock the hydraulics must try to isolate the hydraulic system from shock associated with decompression. The shock can effect the lines and fittings if they are not properly constructed and the hydraulic system does not contain an anti-shock feature.

The Future for Hydraulic Presses

If hydraulic presses continue to improve there is no doubt that they will play an increasingly significant role in industrial production.

1. Europe has already undergone this change. Hydraulic presses enjoy far more widespread use there than in the United States. In that case stringent safety regulations have accelerated the adoption of hydraulic presses.

2. Certainly hydraulic press speeds will be increasing over the next few years. American hydraulic valve manufacturers may soon make available new valves with higher flow capacities and faster response time. But unless a radically different hydraulic circuit design is developed it is unrealistic to predict that hydraulic press speeds will overtake the mechanical press.

3. A change that is already taking place is improved self-powered feeding mechanisms. The future will demand more in this area.

4. Improvements in the anti-shock after breakthrough are already developing. At least two manufacturers have brought out significant new systems to cope with this problem.

5. <u>Activation on withdrawal</u> through electronic light curtains seems to be highly likely to come about. It is not legal to do so on mechanical presses at this time. But experiments being conducted now show this method to be a very promising and safe technique for the freeing of the operator from the task of depressing activating buttons.

6. <u>Programmable Controller.</u> The press sequence can be programmed for a particular job number. The program might set the pressure, stroke length, dwell time, pull-out pressure, and other auxiliary functions and the sequence of these functions. These systems are just now appearing.

7. <u>Human Engineering.</u> Work will continue on hydraulic presses (and on all presses) to improve the comfort and safety of presses. In this area better illumination; quicker machines; comfortable work position; and provisions for easy adjustments are the improvements to be anticipated.

Summary

A hydraulic press can be a useful and productive part of a manufacturing system. The evidence shows increasing popularity of this machine in production applications. The proper selection and use of the machine can be enhanced by a greater understanding of the characteristics of a hydraulic press. For the manufacturing engineer the press however is only one part of a total system which includes tooling, feeding, personal protection, and part handling.

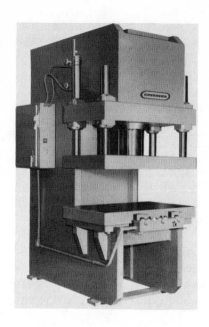

Photo A: 200 ton Open Gap Hydraulic Press with 4-Post Guided Platen.
Bed size 50" R-L x 29" F-B

Photo used for illustration of press only. Does not include guarding.

Photo B: 300 ton 4-Post Hydraulic Press. Bed Size 60" R-L x 57" F-B
Guards at sides removed. For illustration of press only.

Photo C: 200 ton Open Gap Hydraulic Press with cradle for operation as an OBI.

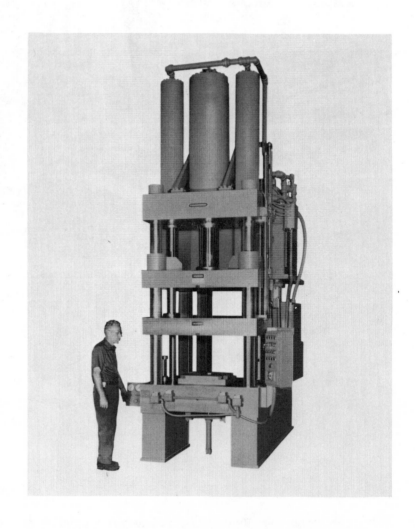

Photo D: 150 ton 4-Post Double Acting Hydraulic Press for deep drawing operations.

APPENDIX

KNOW YOUR PRESSES

Know Your Presses

A. Types of Presses
    1. OBI and OBG
    2. Gap
    3. Single Plate Frame
    4. Solid Frame
    5. Tie Rod Frame
    6. Special Frames
    7. Hydraulic Presses
    8. Press Brake and Gang Punch
    9. Double Action

B. Press Nomenclature

C. Types of Drives
    1. Crank
    2. Eccentric
    3. Eccentric Crank
        (a) Types of Crank Drives
             1. Flywheel
             2. Single Gear
             3. Dougle Gear
             4. Single Gear - Twin Drive
             5. Double Crank - Single Drive
             6. Double Crank - Twin Drive
        (b) Types of Eccentric Drive
             1. Single Point - Single Drive
             2. Single Point - Twin Drive
             3. Two Point - Single Drive
             4. Four Point - Twin Drive
             5. Four Point - Quadruple Drive

D. Press Capacity
    1. Source of Energy
    2. Tonnage Rating
    3. Tonnage Chart

E. Press Equipment
    1. Clutches and Types
        (a) Interconnected
        (b) Low Inertia Clutch
        (c) Low Inertia Brake

2. Counterbalance
   (a) Function
   (b) Adjustment
3. Gibbing
   (a) Types
   (b) Adjustment

F. Cushions
   1. Types of Cushions
   2. Adjustment of Cushions
   3. Loading
   4. Cushion Locks

G. Slide Adjustment
   1. Connection
   2. Hand
   3. Powered
   4. Maximum and Minimum Adjustment Switches
   5. Limits of Elevating Motor

H. Tie Rods
   1. Purpose
   2. Methods of Shrinking
      (a) Torch
      (b) Calrod
      (c) Hydraulic

I. Overloads

J. Electrical Schematic

K. Pneumatic Schematic

L. Lube Schematic

M. Typical Crown Construction

N. Typical Slide Construction

O. Typical Clutch Installation

P. Press Selection Prior to Setting Dies
   1. Selection of Press Based on
      (a) Tonnage Requirement
         1. Rating of Press
         2. Distance up on stroke work is to be contacted
      (b) Size of Die Space Area
      (c) Stroke and Shut Height
      (d) Speed of Press
   2. Location of Dies in Press
      (a) Effects of Improper Loading
      (b) Effects of Improper Press Selection

Q. Suggested Die Setting Procedures
   1. Safety
   2. Installation and Setting

R. Operator Safety

S. Equipment Care

T. Safety Manual
   1. Management Responsibility
   2. Process Planning
   3. Die Setting
   4. Operator Safety
   5. Equipment Care
   6. Safety Tips
   7. Press Warning Signs

TYPES OF FRAMES

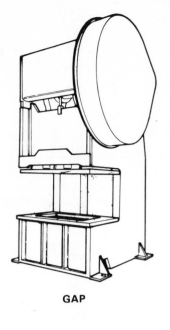

GAP

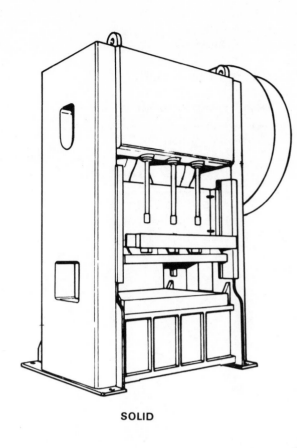

SOLID

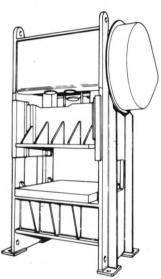

SINGLE PLATE

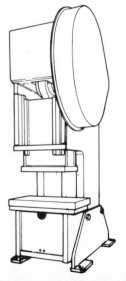

OPEN BACK INCLINABLE

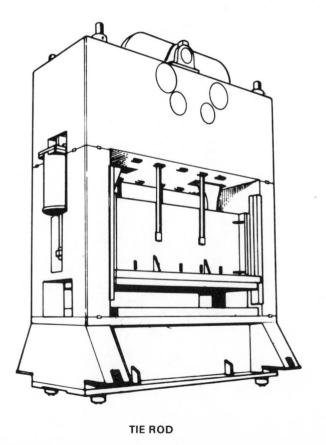

TIE ROD

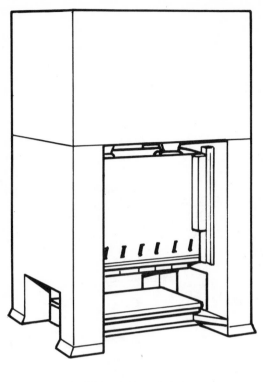

2 COLUMN PRESS

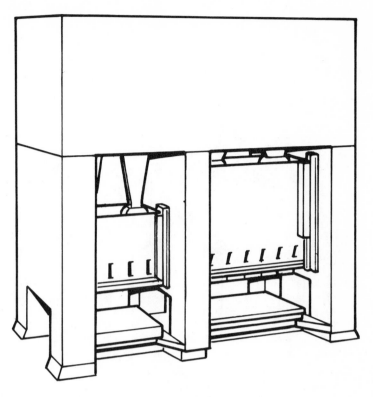

3 COLUMN PRESS

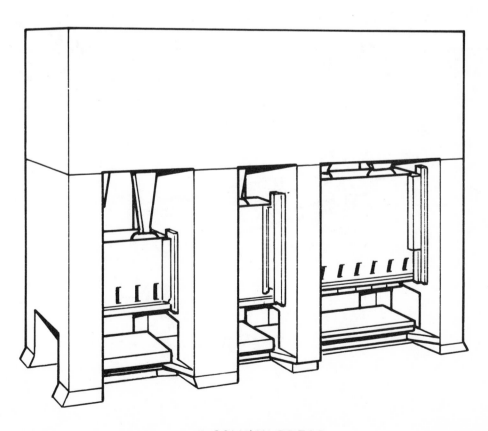

4 COLUMN PRESS

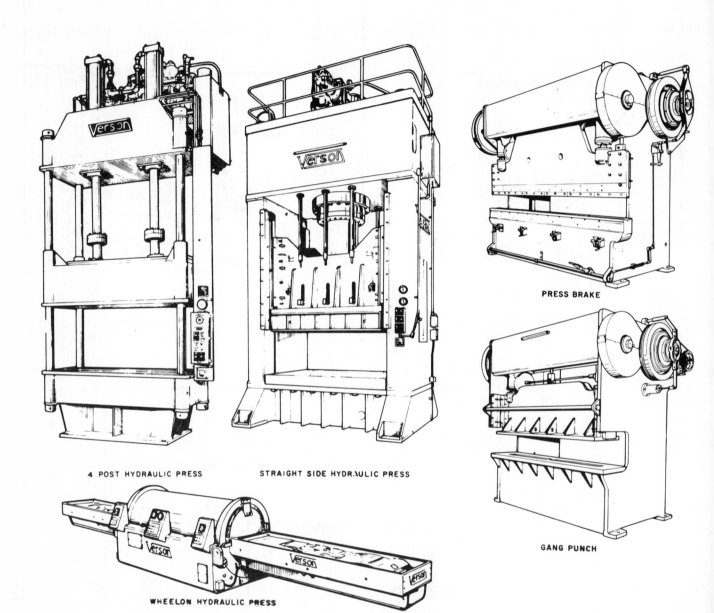

4 POST HYDRAULIC PRESS

STRAIGHT SIDE HYDRAULIC PRESS

PRESS BRAKE

GANG PUNCH

WHEELON HYDRAULIC PRESS

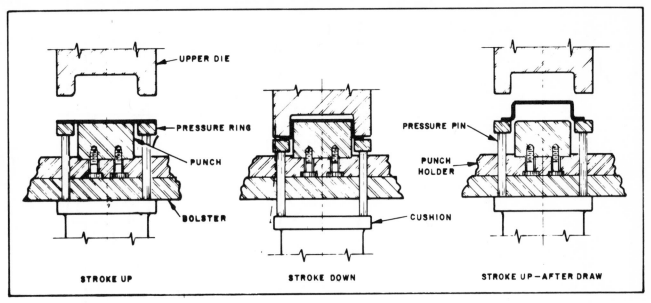

SINGLE-ACTION PRESS

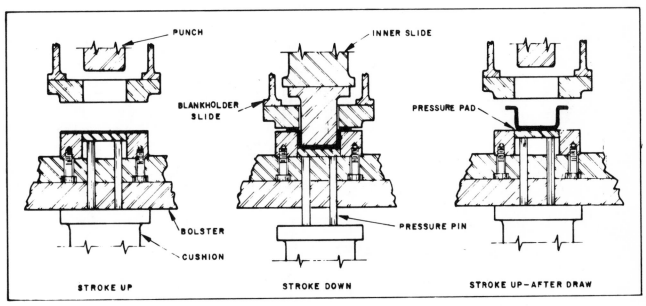

DOUBLE-ACTION PRESS

TYPES OF DRIVES

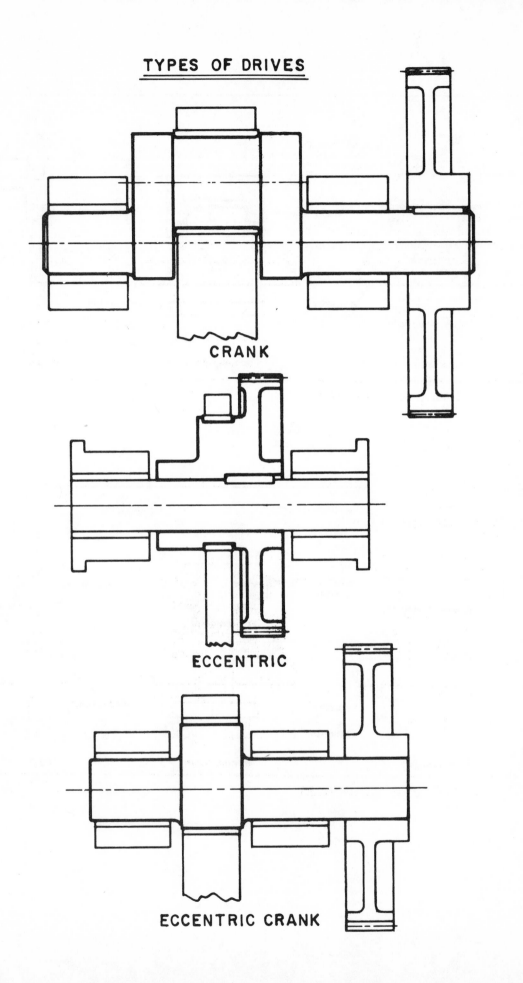

CRANK

ECCENTRIC

ECCENTRIC CRANK

TYPES OF PRESS DRIVES

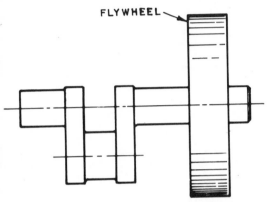

FLYWHEEL OR NON-GEARED DRIVE

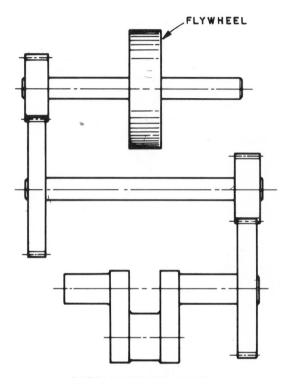

DOUBLE GEARED DRIVE

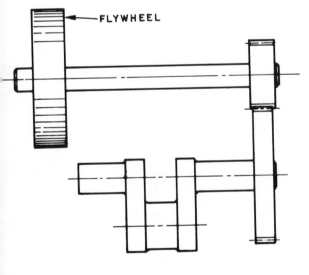

SINGLE GEARED DRIVE

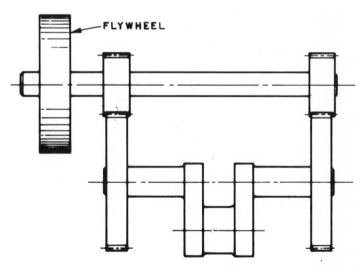

SINGLE GEARED - TWIN DRIVE

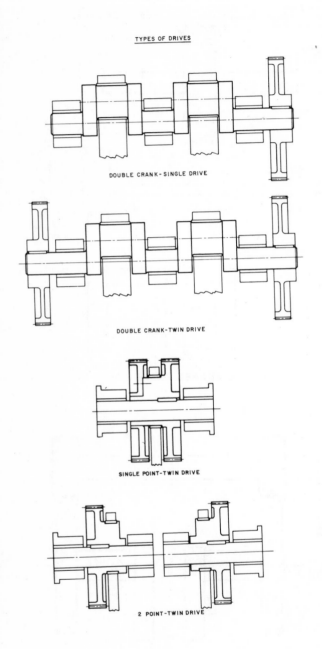

TYPES OF DRIVES

DOUBLE CRANK-SINGLE DRIVE

DOUBLE CRANK-TWIN DRIVE

SINGLE POINT-TWIN DRIVE

2 POINT-TWIN DRIVE

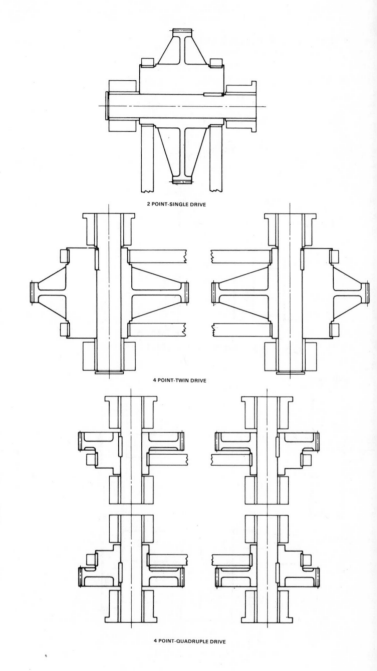

2 POINT-SINGLE DRIVE

4 POINT-TWIN DRIVE

4 POINT-QUADRUPLE DRIVE

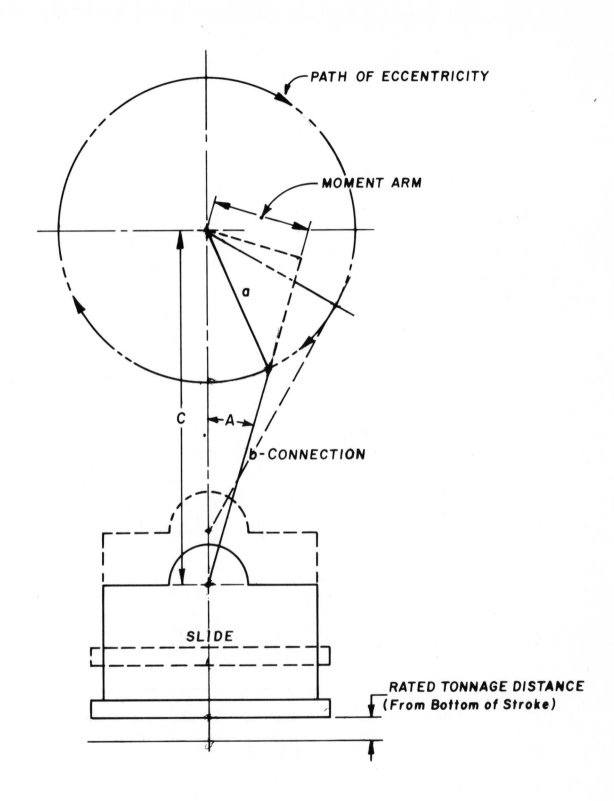

PATH OF ECCENTRICITY

MOMENT ARM

a

C

A

b-CONNECTION

SLIDE

RATED TONNAGE DISTANCE
(From Bottom of Stroke)

ECCENTRIC DRIVE

(calculating total torque to main gears)

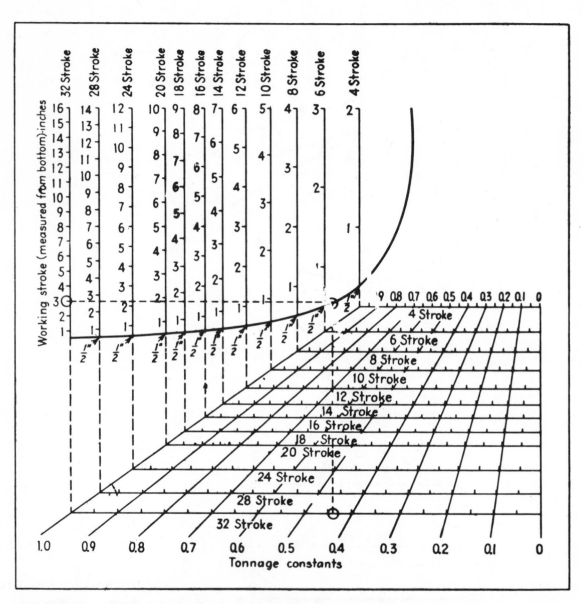

FIG. 1-7. Tonnage stroke relationship curve for obtaining pressure press can exert at any point on the stroke. Simply multiply the rated tonnage capacity of the press by the tonnage constant obtained from the curve.

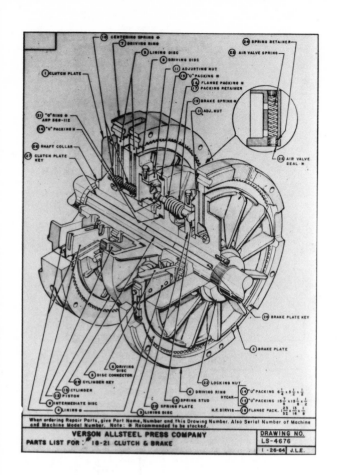

Drawing LS-4676 — Parts list labels:

- 16 CENTERING SPRING *
- 7 DRIVING RING
- 5 LINING DISC
- 3 DRIVING DISC
- 11 ADJUSTING NUT
- 13 "U" PACKING *
- 4 FLANGE PACKING *
- 5 PACKING RETAINER
- 24 SPRING RETAINER
- 1 CLUTCH PLATE
- 23 BRAKE SPRING *
- 11 ADJ. NUT
- 21 "O" RING * ARP 568-112
- 14 "U" PACKING H
- 20 SHAFT COLLAR
- 25 AIR VALVE SPRING
- 27 CLUTCH PLATE KEY
- 25 AIR VALVE SEAL *
- 26 BRAKE PLATE KEY
- 2 BRAKE PLATE
- 6 DRIVING DISC CONNECTOR
- 29 CYLINDER KEY
- 19 CYLINDER
- 18 PISTON
- 17 INTERMEDIATE DISC
- 4 LINING *
- 22 LOCKING NUT
- 7 DRIVING RING HYCAR
- 9 SPRING STUD
- 10 SPRING PLATE
- 5 LINING DISC
- 14 "U" PACKING 6 1/8 x 5 1/2 x 1/2
- 13 "U" PACKING 10 7/8 x 10 1/2 x 1/2
- H.F. SIRVI8 8 FLANGE PACK. 23/16 19/16 1/4

When ordering Repair Parts, give Part Name, Number and this Drawing Number. Also Serial Number of Machine and Machine Model Number. Note: * Recommended to be stocked.

VERSON ALLSTEEL PRESS COMPANY
PARTS LIST FOR: 18-21 CLUTCH & BRAKE

DRAWING NO. LS-4676
1-26-64 J.L.E.

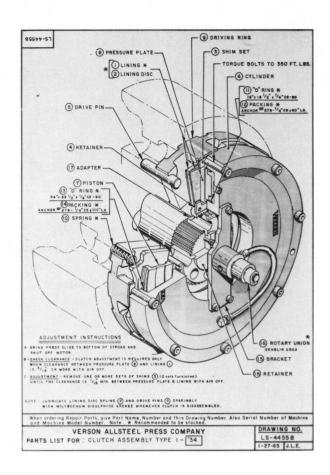

Drawing LS-4455B — Parts list labels:

- 9 DRIVING RING
- 3 SHIM SET
- TORQUE BOLTS TO 350 FT. LBS.
- 8 PRESSURE PLATE
- 1 LINING *
- 2 LINING DISC
- 6 CYLINDER
- 11 "O" RING * 18" x 18 1/2 x 1/4" CS-90
- 12 PACKING * ANCHOR * 276-1/4" CS x 40" LG.
- 5 DRIVE PIN
- 4 RETAINER
- 17 ADAPTER
- 7 PISTON
- 13 "O" RING * 34" x 33 1/2 x 1/4" CS-90
- 14 PACKING * ANCHOR * 276 - 1/4" CS x 110"LG.
- 10 SPRING *
- 16 ROTARY UNION DEUBLIN 250A
- 15 BRACKET
- 19 RETAINER

ADJUSTMENT INSTRUCTIONS

A. BRING PRESS SLIDE TO BOTTOM OF STROKE AND SHUT OFF MOTOR.

B. CHECK CLEARANCE - CLUTCH ADJUSTMENT IS REQUIRED ONLY WHEN CLEARANCE BETWEEN PRESSURE PLATE 8 AND LINING 1 IS 3/16 OR MORE WITH AIR OFF.

C. ADJUSTMENT - REMOVE ONE OR MORE SETS OF SHIMS 3 (2 sets furnished) UNTIL THE CLEARANCE IS 1/16 MIN. BETWEEN PRESSURE PLATE & LINING WITH AIR OFF.

NOTE: LUBRICATE LINING DISC SPLINE 2 AND DRIVE PINS 5 SPARINGLY WITH MOLYBDENUM DISULPHIDE GREASE WHENEVER CLUTCH IS DISASSEMBLED.

When ordering Repair Parts, give Part Name, Number and this Drawing Number. Also Serial Number of Machine and Machine Model Number. Note: * Recommended to be stocked.

VERSON ALLSTEEL PRESS COMPANY
PARTS LIST FOR: CLUTCH ASSEMBLY TYPE I - 34

DRAWING NO. LS-4455B
1-27-65 J.L.E.

VERSON MINIMUM INERTIA AIR COOLED BRAKE

1. BRAKE IS SPRING SET AND AIR RELEASED.
2. SHIM TYPE ADJUSTMENT TO COMPENSATE FOR WEAR.
3. BRAKE IS DESIGNED FOR GENEROUS AIR PASSAGES FOR MAXIMUM COOLING.
4. CAN BE PROVIDED WITH FORCED VENTILATION FOR HEAVY APPLICATIONS.

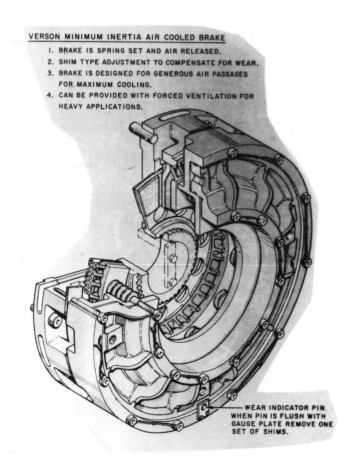

WEAR INDICATOR PIN. WHEN PIN IS FLUSH WITH GAUGE PLATE REMOVE ONE SET OF SHIMS.

COUNTERBALANCES

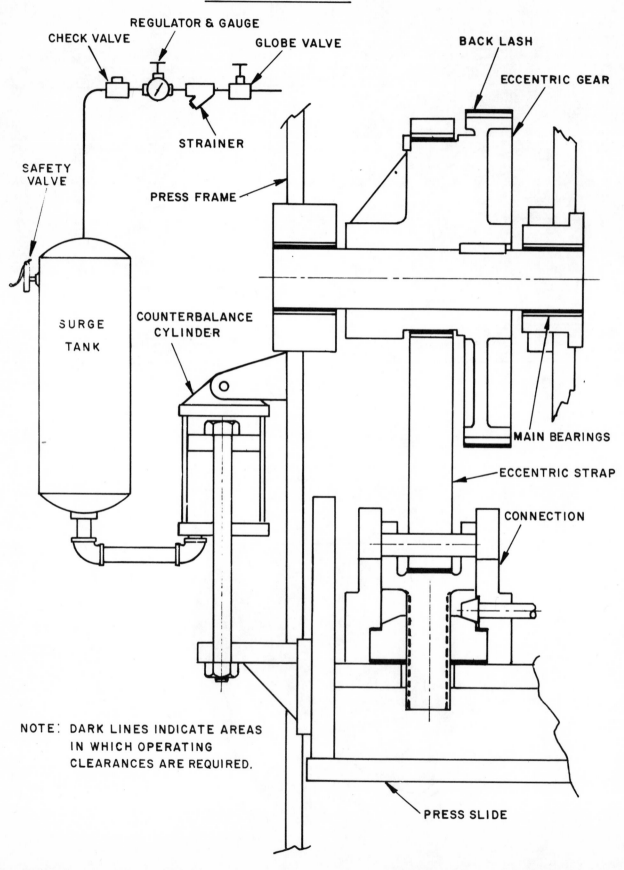

CHECK VALVE

REGULATOR & GAUGE

GLOBE VALVE

STRAINER

BACK LASH

ECCENTRIC GEAR

SAFETY VALVE

PRESS FRAME

SURGE TANK

COUNTERBALANCE CYLINDER

MAIN BEARINGS

ECCENTRIC STRAP

CONNECTION

NOTE: DARK LINES INDICATE AREAS IN WHICH OPERATING CLEARANCES ARE REQUIRED.

PRESS SLIDE

GIBBING

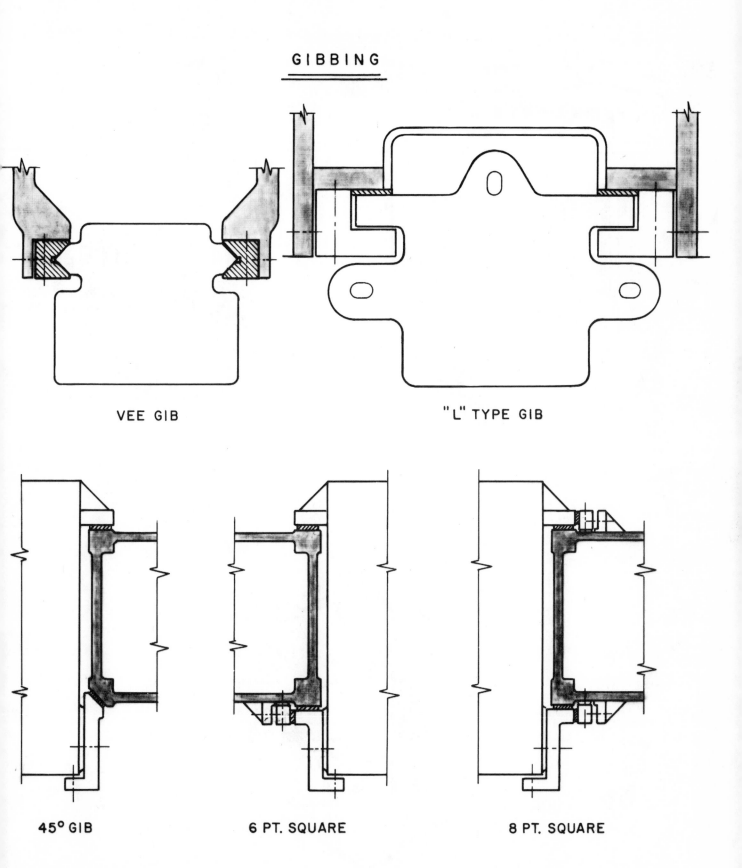

VEE GIB

"L" TYPE GIB

45° GIB

6 PT. SQUARE

8 PT. SQUARE

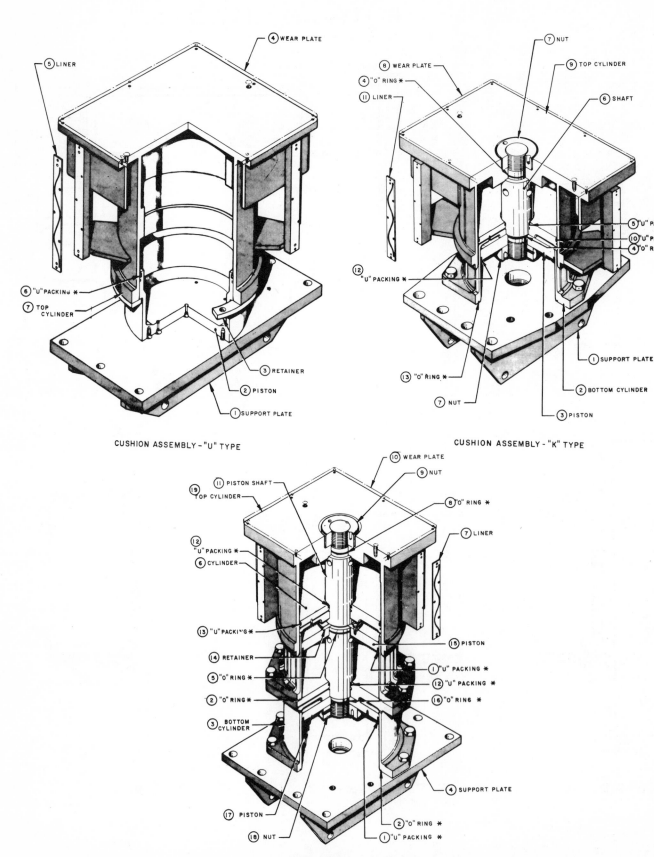

CUSHION ASSEMBLY - "U" TYPE

CUSHION ASSEMBLY - "K" TYPE

CUSHION ASSEMBLY - TYPE KW

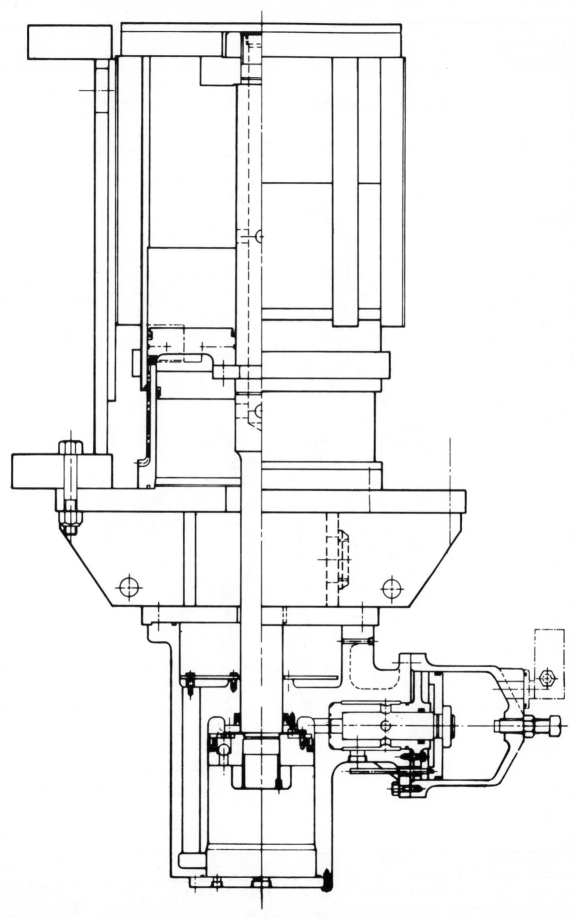

CUSHION WITH LOCK & PULL DOWN FEATURE

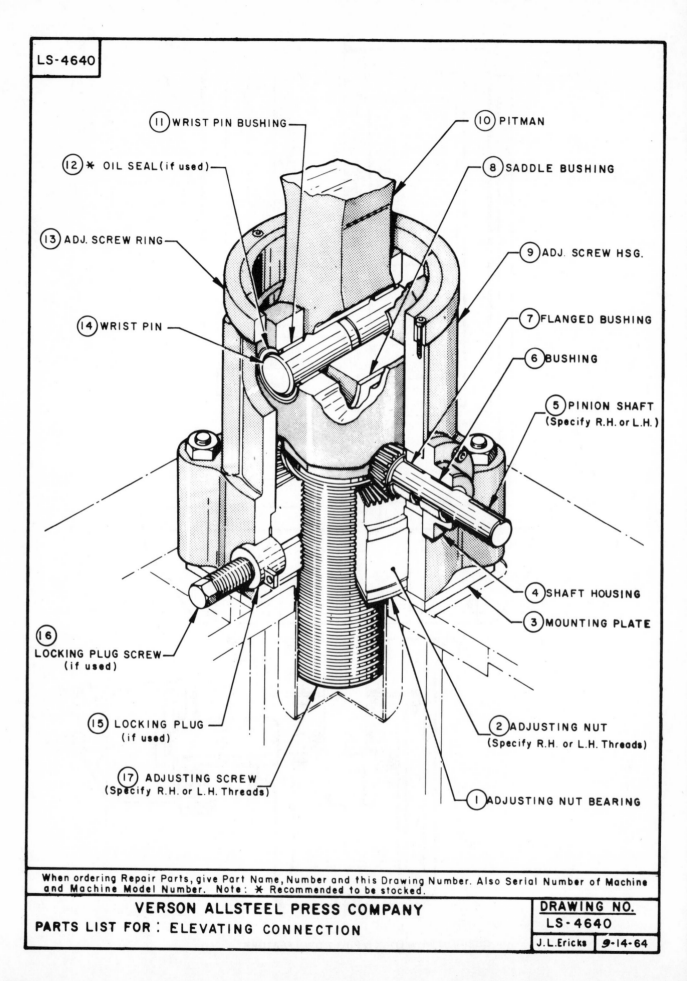

LS-4640

11 WRIST PIN BUSHING

12 ✱ OIL SEAL (if used)

13 ADJ. SCREW RING

14 WRIST PIN

16 LOCKING PLUG SCREW (if used)

15 LOCKING PLUG (if used)

17 ADJUSTING SCREW (Specify R.H. or L.H. Threads)

10 PITMAN

8 SADDLE BUSHING

9 ADJ. SCREW HSG.

7 FLANGED BUSHING

6 BUSHING

5 PINION SHAFT (Specify R.H. or L.H.)

4 SHAFT HOUSING

3 MOUNTING PLATE

2 ADJUSTING NUT (Specify R.H. or L.H. Threads)

1 ADJUSTING NUT BEARING

When ordering Repair Parts, give Part Name, Number and this Drawing Number. Also Serial Number of Machine and Machine Model Number. Note: ✱ Recommended to be stocked.

VERSON ALLSTEEL PRESS COMPANY
PARTS LIST FOR: ELEVATING CONNECTION

DRAWING NO.
LS-4640

J.L.Ericks   9-14-64

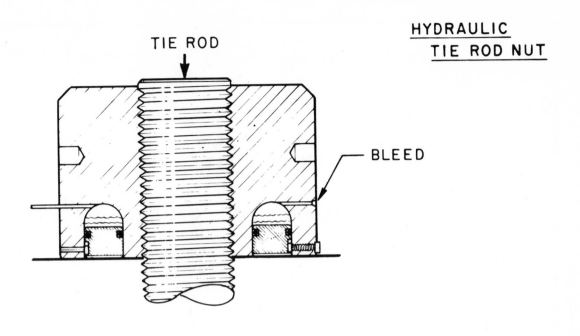

TIE ROD

BLEED

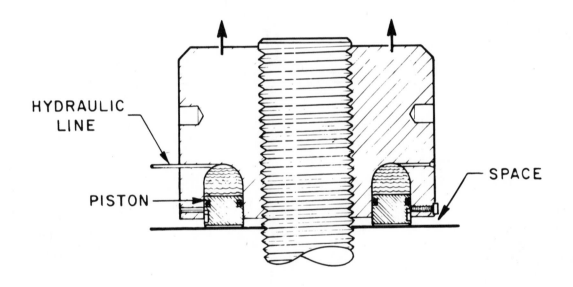

HYDRAULIC
LINE

PISTON

SPACE

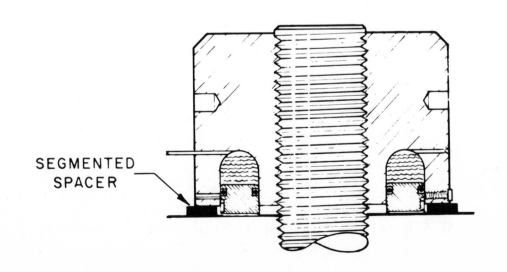

SEGMENTED
SPACER

TOP OF PRESS CROWN
Showing
HYDRAULIC TIE ROD PIPING TO HYDRAULIC PUMP

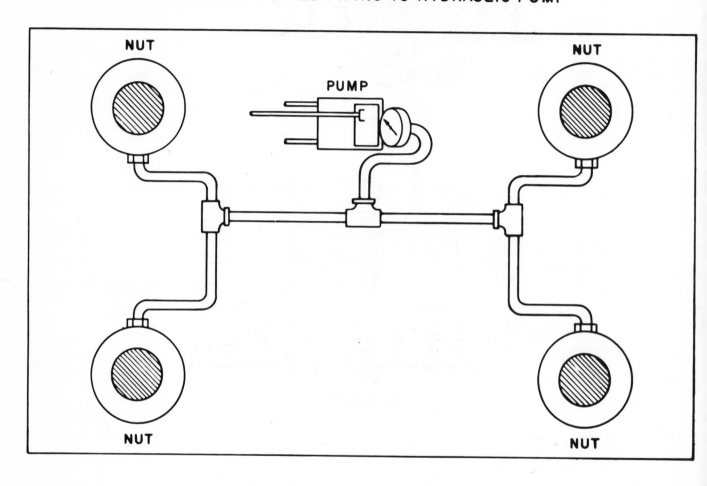

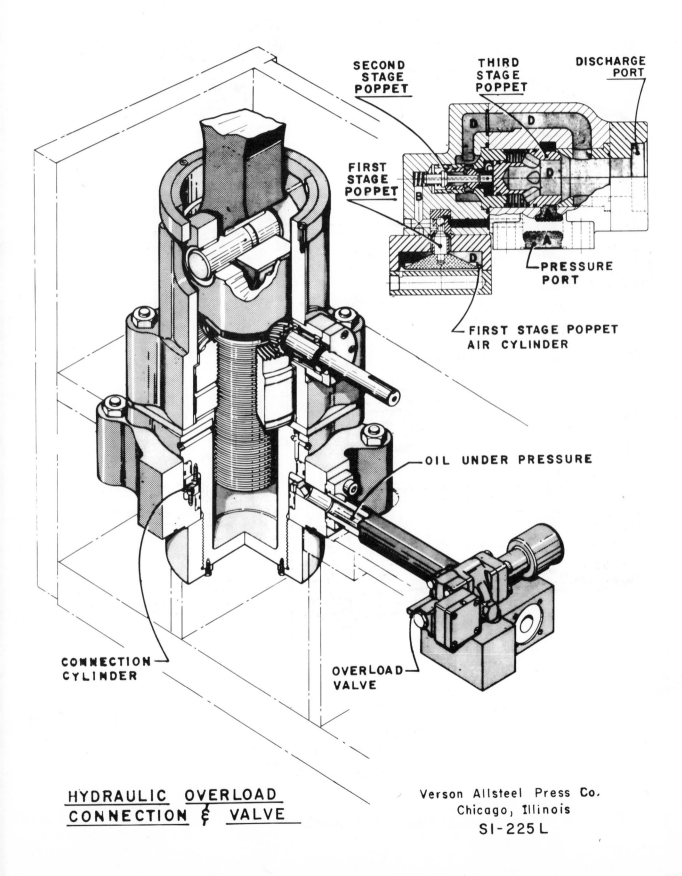

SECOND STAGE POPPET

THIRD STAGE POPPET

DISCHARGE PORT

FIRST STAGE POPPET

PRESSURE PORT

FIRST STAGE POPPET AIR CYLINDER

OIL UNDER PRESSURE

CONNECTION CYLINDER

OVERLOAD VALVE

HYDRAULIC OVERLOAD
CONNECTION & VALVE

Verson Allsteel Press Co.
Chicago, Illinois
SI-225 L

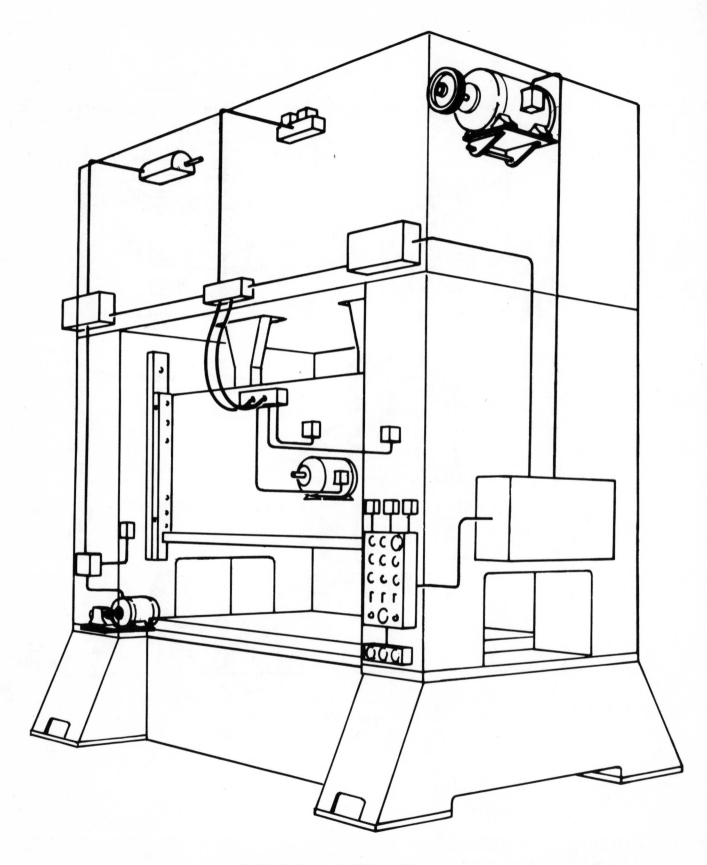

ELECTRICAL SCHEMATIC

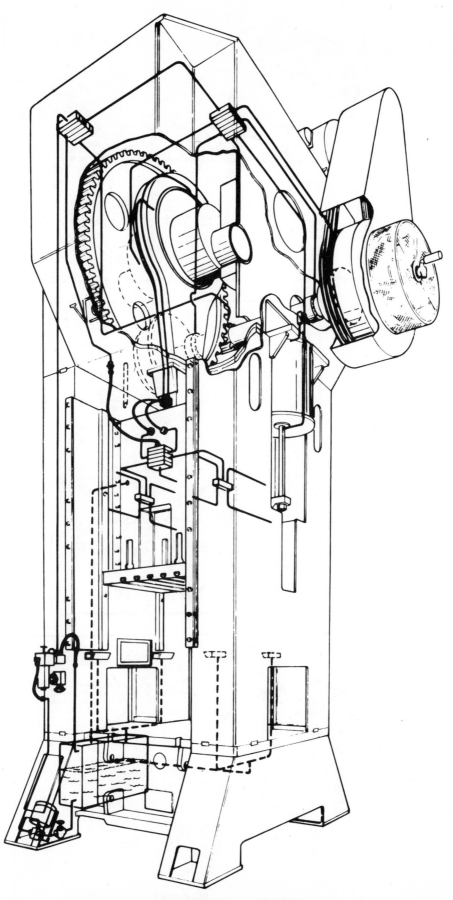

CASCADE LUBRICATION SYSTEM

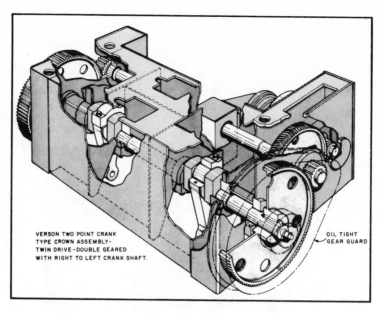

VERSON TWO POINT CRANK
TYPE CROWN ASSEMBLY-
TWIN DRIVE-DOUBLE GEARED
WITH RIGHT TO LEFT CRANK SHAFT.

OIL TIGHT
GEAR GUARD

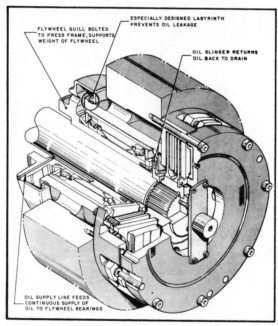

FLYWHEEL QUILL BOLTED
TO PRESS FRAME, SUPPORTS
WEIGHT OF FLYWHEEL

ESPECIALLY DESIGNED LABYRINTH
PREVENTS OIL LEAKAGE

OIL SLINGER RETURNS
OIL BACK TO DRAIN

OIL SUPPLY LINE FEEDS
CONTINUOUS SUPPLY OF
OIL TO FLYWHEEL BEARINGS.

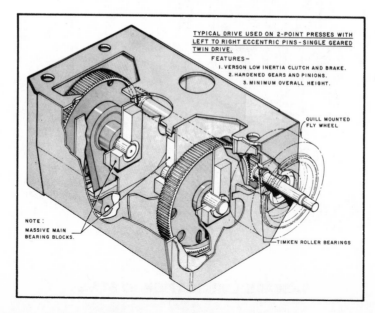

TYPICAL DRIVE USED ON 2-POINT PRESSES WITH
LEFT TO RIGHT ECCENTRIC PINS-SINGLE GEARED
TWIN DRIVE.

FEATURES-
1. VERSON LOW INERTIA CLUTCH AND BRAKE.
2. HARDENED GEARS AND PINIONS.
3. MINIMUM OVERALL HEIGHT.

QUILL MOUNTED
FLY WHEEL

NOTE:
MASSIVE MAIN
BEARING BLOCKS.

TIMKEN ROLLER BEARINGS

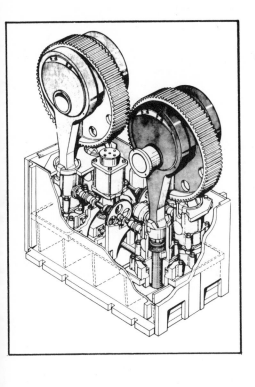

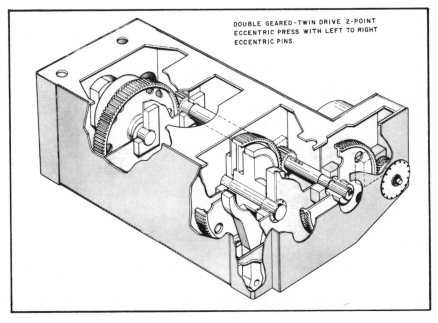

DOUBLE GEARED-TWIN DRIVE 2-POINT
ECCENTRIC PRESS WITH LEFT TO RIGHT
ECCENTRIC PINS.

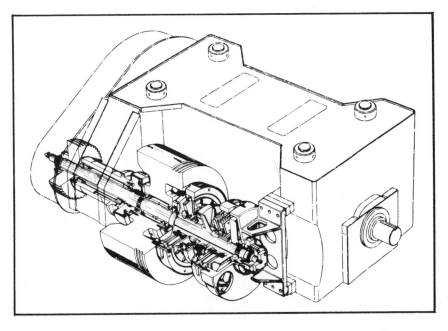

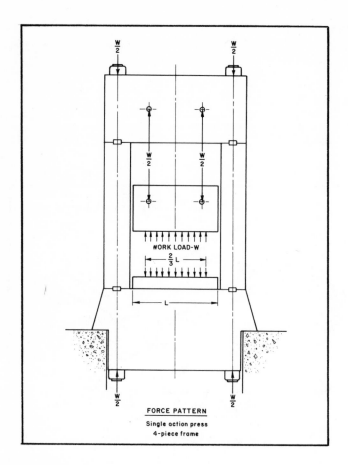

FORCE PATTERN

Single action press
4-piece frame

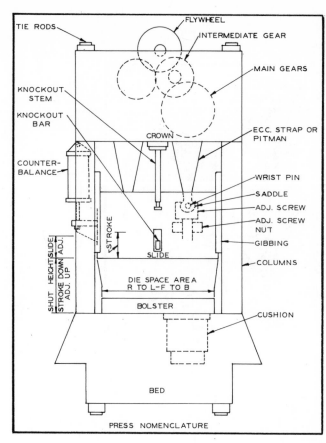

PRESS NOMENCLATURE

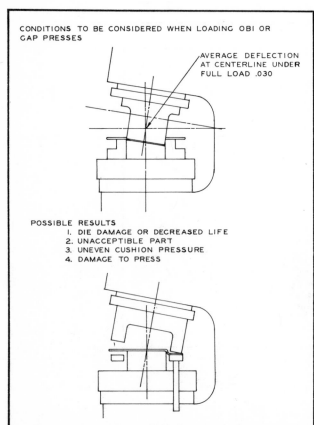

CONDITIONS TO BE CONSIDERED WHEN LOADING OBI OR GAP PRESSES

AVERAGE DEFLECTION AT CENTERLINE UNDER FULL LOAD .030

POSSIBLE RESULTS
1. DIE DAMAGE OR DECREASED LIFE
2. UNACCEPTIBLE PART
3. UNEVEN CUSHION PRESSURE
4. DAMAGE TO PRESS

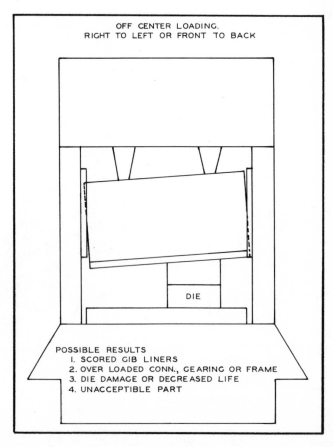

OFF CENTER LOADING.
RIGHT TO LEFT OR FRONT TO BACK

DIE

POSSIBLE RESULTS
1. SCORED GIB LINERS
2. OVER LOADED CONN., GEARING OR FRAME
3. DIE DAMAGE OR DECREASED LIFE
4. UNACCEPTABLE PART

INDEX